Ali AL-Hamdani
Alaa Al-Jizany
Hayfa Rashid

Microlens Arrays Solar Concentrator Design using ZEMAX

Ali AL-Hamdani
Alaa Al-Jizany
Hayfa Rashid

Microlens Arrays Solar Concentrator Design using ZEMAX

LAP LAMBERT Academic Publishing

Impressum / Imprint

Bibliografische Information der Deutschen Nationalbibliothek: Die Deutsche Nationalbibliothek verzeichnet diese Publikation in der Deutschen Nationalbibliografie; detaillierte bibliografische Daten sind im Internet über http://dnb.d-nb.de abrufbar.

Alle in diesem Buch genannten Marken und Produktnamen unterliegen warenzeichen-, marken- oder patentrechtlichem Schutz bzw. sind Warenzeichen oder eingetragene Warenzeichen der jeweiligen Inhaber. Die Wiedergabe von Marken, Produktnamen, Gebrauchsnamen, Handelsnamen, Warenbezeichnungen u.s.w. in diesem Werk berechtigt auch ohne besondere Kennzeichnung nicht zu der Annahme, dass solche Namen im Sinne der Warenzeichen- und Markenschutzgesetzgebung als frei zu betrachten wären und daher von jedermann benutzt werden dürften.

Bibliographic information published by the Deutsche Nationalbibliothek: The Deutsche Nationalbibliothek lists this publication in the Deutsche Nationalbibliografie; detailed bibliographic data are available in the Internet at http://dnb.d-nb.de.

Any brand names and product names mentioned in this book are subject to trademark, brand or patent protection and are trademarks or registered trademarks of their respective holders. The use of brand names, product names, common names, trade names, product descriptions etc. even without a particular marking in this work is in no way to be construed to mean that such names may be regarded as unrestricted in respect of trademark and brand protection legislation and could thus be used by anyone.

Coverbild / Cover image: www.ingimage.com

Verlag / Publisher:
LAP LAMBERT Academic Publishing
ist ein Imprint der / is a trademark of
OmniScriptum GmbH & Co. KG
Heinrich-Böcking-Str. 6-8, 66121 Saarbrücken, Deutschland / Germany
Email: info@lap-publishing.com

Herstellung: siehe letzte Seite /
Printed at: see last page
ISBN: 978-3-659-80122-8

Copyright © 2015 OmniScriptum GmbH & Co. KG
Alle Rechte vorbehalten. / All rights reserved. Saarbrücken 2015

Microlens Arrays Solar Concentrator Design using ZEMAX

Ali H. Al-Hamdani

Alaa Badr Hasan Al-Jizany

Hayfa Ghazi Rashid

List of Symbols

Symbol	Meaning	Unit
A	Incidence Angles	Degree
$ά$	Absorption Coefficient of Glass	cm^{-1}
θ	Acceptance Angle	Degree
θ_I	Ideal Acceptance Angle	Degree
θ_R	Real Acceptance Angle	Degree
θ_S	Angular Aperture	Degree
Λ	Wavelength	µm
H	Optical Efficiency	-
C	Atmospheric Extinction Coefficient	-
C_{geo}	Geometric Concentration Ratio	-
C_{flux}	Flux Concentration	-
C_{lens}	Geometric Concentration of Lens	-
C_{max}	Maximum Concentration	-
D	Diameter of the Lens	Mm
D	Resolvable Distance	Mm
d_n	day number	-
E_{ext}	Extraterrestrial Solar Illuminance	W/m^2
E_{sc}	Solar Illuminance Constant	W/m^2
E_{dn}	Direct Normal Illuminance	W/m^2
F	Focal length	Mm
$F/\#$	Focal Number	-
N	Refractive Index	-
Th	Thickness	Mm
T	Transmittance	-

List of Abbreviations

Abbreviation	Meaning
AM	Relative Optical Air Mass
AU	Astronomical Unit
BK7	Borosilicate Glass
CAP	Concentration Acceptance Product
CSP	Concentrated Solar Power
CPV	Concentration Photovoltaic
CdTe	Cadmium Telluride
CIS	Copper Indium Selenide
CCD	Charge-Coupled Device
CTE	Coefficients of Thermal Expansion
DOD	Drop-on-Demand
DRPS	Double side Ray Propagation Sample
EMR	Electromagnetic Radiation
FDTD	Finite-Difference Time-Domain
GEO	Geometrical Spot Radius
GRIN	Gradient-Index Lenses
IR	Infrared
LCD	Liquid Crystal Display
LDE	Lens Data Editor
LED	Light Emitting Diode
MLA	Microlens Arrays
MTF	Modulation Transfer Functions
NSC	Non Sequential Component
NSOM	Near-field Scanning Optical Microscopy
NA	Numerical Aperture
OPD	Optical Path Difference
OTF	Optical Transfer Function
PMMA	Poly-Methyl Methacrylate

Abbreviation	Meaning
PSF	Point Spread Function
RIE	Reactive Ion Etching
PVC	Photovoltaic Cell
RMS	Root Mean Square
a-Si	Amorphous Silicon
c-Si	Monocrystalline Silicon
nc-Si	Nanocrystals Silicon
SC	Sequential Component
SEGS	Solar Energy Generating Systems
SLA	Stretched Lens Array
SPAD	Single Photon Avalanche Detector
SPM	Scanning Probe Microscopy
SRPS	Single side Ray Propagation Sample
TIR	Total Internal Reflection
UVA	Ultraviolet class A
UVB	Ultraviolet class B

CONTENTS

Title	Page No.
List of symbols	II
List of abbreviations	III
Contents	V
Chapter one: Introduction	1
1.1 Renewable Energy	2
1.2 Solar Radiation	3
1.3 Solar Concentrators	6
1.3.1 Parabolic Trough	6
1.3.2 Fresnel Reflectors	7
1.3.3 Dish Stirling	7
1.3.4 Solar Power Tower	9
1.4 Concentrated photovoltaic	9
1.5 Solar Photovoltaic	10
1.6 Acceptance angle	12
Chapter two: Micro Optics	15
2.1 Introduction	16
2.2 Microlens Arrays	16
2.3 Micro Optics in Nature	18
2.3.1 Single-Aperture Eyes	18
2.3.2 Compound Eyes	19
2.3.2.1 Apposition and Super-Position	19
2.4 Nano-Optics	20
2.5 Optical Systems	21
2.5.1 Imaging Optical Systems	21
2.5.1.1 Diffraction	21
2.5.1.2 Aberration	22
2.5.2 Non-Imaging Optical Systems	23
2.5.2.1 Characteristics of non-imaging optical system	23
2.6 Computer Performance Evaluation	24
2.6.1 Optical Resolution	25
2.6.2 Transverse Ray Aberrations Curves	25
2.6.3 Optical Path Difference	27
2.6.4 Optical Transfer Function	28
2.6.5 Spot Diagrams	29

Title	Page No.
2.6.6 Encircled Energy	29
Chapter three: Optical Designs	31
3.1 Introduction	32
3.2 Optical Software Programs	32
3.3 ZEMAX Software Program	33
3.3.1 Optimization	33
3.3.2 Merit Function	34
3.4 Ray-Tracing Mode	34
3.5 Sequential Ray-Tracing Mode in ZEMAX	35
3.6 Non-Sequential Ray-Tracing Mode in ZEMAX	37
3.7 Optical System prototypes that Used in ZEMAX	38
3.7.1 Plano-Convex Microlens Arrays (1-D)	39
3.7.2 Plano-Convex Microlens Arrays (2-D)	40
3.7.3 Plano-convex microlens arrays with slab waveguide	40
3.8 Optical Materials	44
3.9 Lens Array	44
3.10 Waveguide Layer	46
3.11 Cladding Layer	46
References	51

Chapter one
Introduction

Chapter one

Introduction

1.1 Renewable Energy

   Renewable energy is the energy which comes from natural resources such as sunlight, wind, rain, tides, waves, biomass and geothermal heat, which are renewable (naturally replenished). About 16% of global final energy consumption comes from renewable. In its various forms, it derives directly from the sun, or from heat generated deep within the earth.[1]
   Renewable energy can be particularly suitable for developing countries. In rural and remote areas, transmission and distribution of energy generated from fossil fuels can be difficult and expensive. Producing renewable energy locally can offer a viable alternative. The incentive to use 100% renewable energy is created by global warming and ecological as well as economic concerns.
   Renewable energy resources and significant opportunities for energy efficiency exist over wide geographical areas, in contrast to other energy sources, which are concentrated in a limited number of countries. Rapid deployment of renewable energy and energy efficiency, and technological diversification of energy sources, would result in significant energy security and economic benefits.[2]

   Solar energy is the most important types of renewable energy which derived from the sun through the form of solar radiation. Solar powered electrical generation relies on photovoltaic and heat engines. A partial list of other solar applications includes space heating and cooling through solar architecture, day lighting, solar hot water, solar cooking, and high temperature process heat for industrial purposes.
   Solar technologies are broadly characterized as either passive solar or active solar depending on the way they capture, convert and distribute solar energy. Active solar techniques include the use of photovoltaic panels and solar thermal collectors to harness the energy. Passive solar techniques include orienting a building to the Sun, selecting materials with favorable thermal mass or light dispersing properties.
   Solar energy production has been increasing by an average of more than 20% each year since 2002, making it a fast-growing energy technology, while wind is often cited as the fastest growing energy source as shown in figure 1.1[3].

   In photovoltaic cells (PVC) the concentrated sunlight is converted directly to electricity via the photoelectric effect. These Concentrating technologies exist in several forms, like microlens arrays system, waveguide system and textured solar cells. [3]

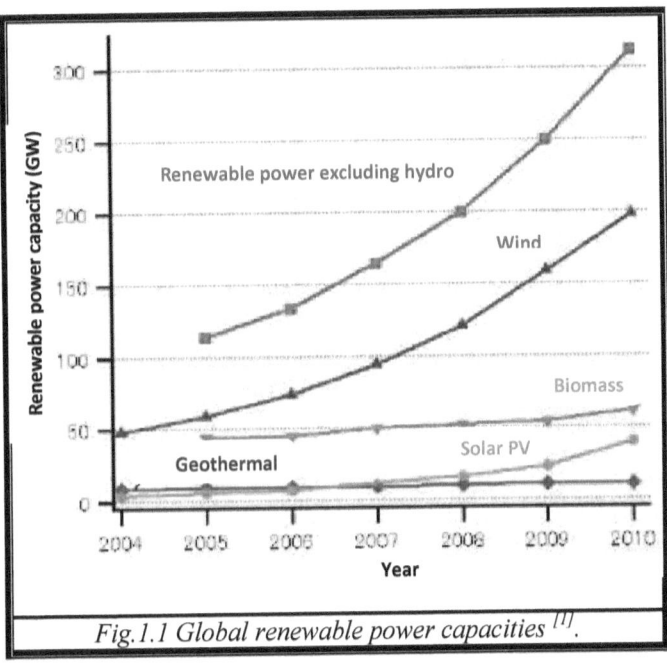

Fig.1.1 Global renewable power capacities [1].

1.2 Solar Radiation

Solar radiation is the total frequency spectrum of electromagnetic radiation (EMR) given off by the Sun, particularly infrared, visible, and ultraviolet light. On Earth, sunlight is filtered through the Earth's atmosphere, and solar radiation is obvious as daylight when the sun is above the horizon [4].

Solar radiation consists of a direct circumsolar component, and a diffuse component. Circumsolar radiation is light coming directly from the solar disk, approximately ±0.25 degrees, plus light that is forward scattered within the atmosphere and exists within a narrow cone around the sun up to ±5 degrees. Diffuse radiation is defined as all light that exists outside the ±5 degree cone [4].

The World Meteorological Organization uses the term "sunshine duration" to mean the cumulative time during which an area receives direct irradiance from the Sun of at least 120 W/m^2 [4].

Direct sunlight has a luminous efficacy of about 93 lum / W of radiant flux. Bright sunlight provides illuminance of approximately 100,000 lux (lum/m^2) at the earth's surface [5].

The spectrum of the sun's solar radiation is close to that of a black body with a temperature of about 5800° K (figure 1.2). The sun emits EMR across most of the electromagnetic spectrum. Although the sun produces gamma rays as a result of the nuclear fusion process, these super high energy photons are converted to lower energy photons before they reach the Sun's surface and are emitted out into space. As a result, the sun doesn't give off any gamma rays. The sun does, however, emit x-rays, ultraviolet, visible light, infrared, and even radio waves. When ultraviolet radiation is not absorbed by the atmosphere or other protective coating, it can cause damage to the skin known as sunburn or trigger an adaptive change in human skin pigmentation [5].

The spectrum of electromagnetic radiation striking the earth's atmosphere spans a range of 100 nm to about 1 mm. This can be divided into five regions in increasing order of wavelengths [6]:

- Ultraviolet C or (UVC) range, which spans a range of 100 to 280 nm. The term ultraviolet refers to the fact that the radiation is at higher frequency than violet light (and, hence also invisible to the human eye). Owing to absorption by the atmosphere very little reaches the earth's surface (Lithosphere). This spectrum of radiation has germicidal properties, and is used in germicidal lamps.
- Ultraviolet B or (UVB) range spans 280 to 315 nm. It is also greatly absorbed by the atmosphere, and along with UVC is responsible for the photochemical reaction leading to the production of the ozone layer.
- Ultraviolet A or (UVA) spans 315 to 400 nm. It has been traditionally held as less damaging to DNA, and hence used in tanning and (Psoralen UVA) therapy for psoriasis.
- Visible range or light spans 380 to 780 nm. As the name suggests, it is this range that is visible to the naked eye.
- Infrared (IR) range that spans 700 nm to 10^6 nm (1 mm). It is responsible for an important part of the electromagnetic radiation that reaches the earth.

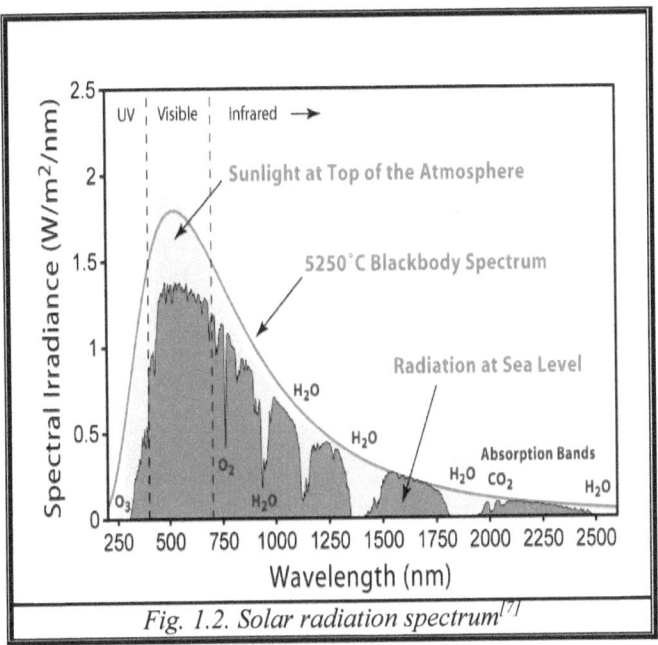

Fig. 1.2. Solar radiation spectrum [7]

To calculate the amount of sunlight reaching the ground, both the elliptical orbit of the Earth and the attenuation by the earth's atmosphere have to be taken into account. The extraterrestrial solar illuminance (E_{ext}), corrected for the elliptical orbit by using the day number of the year (d_n), is given by [8]

$$E_{ext} = E_{sc}\left[1 + 0.033412\ cos\left(2\pi\frac{d_n - 3}{365}\right)\right] \quad (1.1)$$

Where d_n=1 on January 1; d_n=2 on January 2; d_n=32 on February 1, etc. In this formula dn-3 is used, because in modern times earth's perihelion, the closest approach to the Sun and therefore the maximum E_{ext} occurs around January 3 each year. The value of 0.033412 is determined knowing that the difference between the perihelion (0.98328989 AU) and the aphelion (1.01671033 AU) [8].

The solar illuminance constant (E_{sc}), is equal to 128×10^3 lux. The direct normal illuminance (E_{dn}), corrected for the attenuating effects of the atmosphere is [8]:

$$E_{dn} = E_{ext}\ e^{-cAM} \quad (1.2)$$

Where c is the atmospheric extinction coefficient and (AM) is the relative optical air mass that measures of the amount of atmosphere between the observer and the sun. When the sun is directly overhead, the air mass is 1 When the sun is near the horizon, the air mass can be 4 or higher.

The solar constant, a measure of flux density, is the amount of incoming solar electromagnetic radiation per unit area that would be incident on a plane perpendicular to the rays, at a distance of one astronomical unit (AU). The "solar constant" includes all types of solar radiation, not just the visible light. Its average value was thought to be approximately (1.366) kW/m², varying slightly with solar activity, but recent recalibrations of the relevant satellite observations indicate a value closer to 1.361 kW/m² is more realistic.[9]

1.3 Solar Concentrators

Solar concentrator is used to produce electricity (sometimes called solar thermoelectricity, usually generated through steam). Concentrated-solar technology systems use mirrors or lenses with tracking systems to focus a large area of sunlight onto a small area. The concentrated light is then used as light or heat source for a conventional power plant.

Concentrating technologies exist in four common forms, namely parabolic trough, dish stirling, concentrating linear Fresnel reflector, and solar power tower[10]:

1.3.1 Parabolic Trough

A parabolic trough consists of a linear parabolic reflector that concentrates light onto a receiver positioned along the reflector's focal line as shown in figure 1.3. The receiver is a tube positioned directly above the middle of the parabolic mirror and filled with a working fluid. The reflector follows the sun during the daylight hours by tracking along a single axis. A working fluid (e.g. molten salt) is heated to 150–350 °C as it flows through the receiver and is then used as a heat source for a power generation system.[11] Trough systems are the most developed concentration technology.

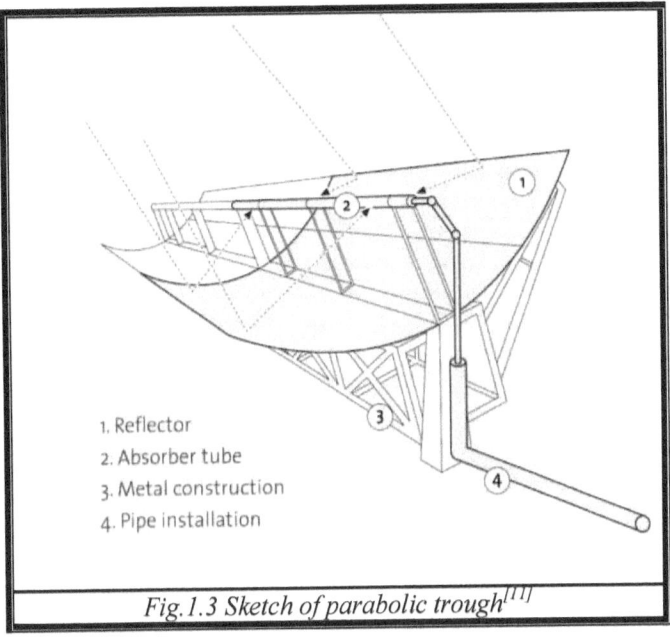

1. Reflector
2. Absorber tube
3. Metal construction
4. Pipe installation

Fig.1.3 Sketch of parabolic trough[11]

1.3.2 Fresnel Reflectors

Fresnel reflectors are made of many thin, flat mirror strips to concentrate sunlight onto tubes through which working fluid is pumped as shown in figure 1.4. Flat mirrors allow more reflective surface in the same amount of space as a parabolic reflector, thus capturing more of the available sunlight, and they are much cheaper than parabolic reflectors[11].

1.3.3 Dish Stirling

A dish stirling or dish engine system consists of a stand-alone parabolic reflector that concentrates light onto a receiver positioned at the reflector's focal point as shown in figure 1.5. The reflector tracks the sun along two axes. The working fluid in the receiver is heated to 250–700 °C and then used by a stirling engine to generate power. Parabolic-dish systems provide the highest solar-to-electric efficiency among concentration technologies. [12]

Fig.1.4 Fresnel reflectors[11]

Fig.1.5 Dish stirling[12]

1.3.4 Solar Power Tower

A solar power tower consists of an array of dual-axis tracking reflectors that concentrate sunlight on a central receiver atop a tower as shown in figure 1.6. The receiver contains a fluid deposit, which can consist of sea water. The working fluid in the receiver is heated to 500–1000 °C and then used as a heat source for a power generation or energy storage system.[12] Power-tower development is less advanced than trough systems, but they offer higher efficiency and better energy storage capability.

Fig.1.6 Solar power tower[12]

1.4 Concentrated photovoltaic

Concentrated photovoltaic (CPV) technology uses optics such as lenses or curved mirrors to concentrate a large amount of sunlight onto a small area of solar photovoltaic cells to generate electricity[13]. Compared to non-concentrated photovoltaic, CPV systems can save money on the cost of the solar cells, since a smaller area of photovoltaic material is required. Because a smaller PV area is required, CPVs can use the more expensive high-efficiency solar cells. To get the sunlight focused on the small PV area, CPV systems require spending extra money on concentrating optics (lenses or mirrors), solar trackers, and cooling systems. Because of these extra costs, CPV is far less common today than non-concentrated photovoltaic. However, ongoing researches and developments are trying to improve CPV technology and lower costs.

CPV also competes with concentrated solar thermal. CPV turns the sunlight directly into electricity, while solar thermal turns the sunlight into heat, and then turns the heat into electricity. Solar thermal is more common than CPV, although the two technologies are sometimes combined.

CPV systems operate most efficiently in concentrated sunlight, as long as the solar cell is kept cool through use of heat sinks. Diffuse light, which occurs in cloudy and overcast conditions, cannot be concentrated. To reach their maximum efficiency, CPV systems must be located in areas that receive plentiful direct sunlight.

The design of photovoltaic concentrators introduces a very specific optical design problem, with features that makes it different from any other optical design. It has to be efficient, suitable for mass production, capable of high concentration, insensitive to manufacturing and mounting inaccuracies, and capable of providing uniform illumination of the cell. All these reasons make nonimaging optics the most suitable for CPV[13].

1.5 Solar Photovoltaic

Photovoltaic is the field of technology and research related to the practical application of photovoltaic cells in producing electricity from light, though it is often used specifically to refer to the generation of electricity from sunlight. Cells can be described as photovoltaic even when the light source is not necessarily sunlight (lamplight, artificial light, etc.). In such cases the cell is sometimes used as a photo-detector (for example infrared detectors), detecting light or other electromagnetic radiation near the visible range, or measuring light intensity [14].

Various materials display varying efficiencies and have varying costs. Materials for efficient solar cells must have characteristics matched to the spectrum of available light. Some cells are designed to efficiently convert wavelengths of solar light that reach the Earth surface. However, some solar cells are optimized for light absorption beyond Earth's atmosphere as well. Light absorbing materials can often be used in multiple physical configurations to take advantage of different light absorption and charge separation mechanisms.

Materials presently used for photovoltaic solar cells include monocrystalline silicon (c-Si), polycrystalline silicon (poly-Si), amorphous silicon (a-Si), cadmium telluride (CdTe), and copper indium selenide (CIS).[14]

Many currently available solar cells are made from bulk materials that are cut into wafers between 180 to 240 μm thick that are then processed like other semiconductors.

Other materials are made as thin-films layers, organic dyes, and organic polymers that are deposited on supporting substrates. A third group are made from nanocrystals (nc-Si) and used as quantum dots (electron-confined nanoparticles). Silicon remains the only material that is well-researched in both bulk and thin-film forms. The most prevalent bulk material for solar cells is crystalline silicon (figure 1.7), also known as "solar grade silicon". [14]

Amorphous silicon has a higher band gap (1.7 eV) than crystalline silicon (c-Si) (1.1 eV) [13], which means it absorbs the visible part of the solar spectrum more strongly than the infrared portion of the spectrum. As nc-Si has about the same band gap as c-Si, the nc-Si and a-Si can advantageously be combined in thin layers, creating a layered cell called a tandem cell. The top cell in a-Si absorbs the visible light and leaves the infrared part of the spectrum for the bottom cell in nc-Si [15].

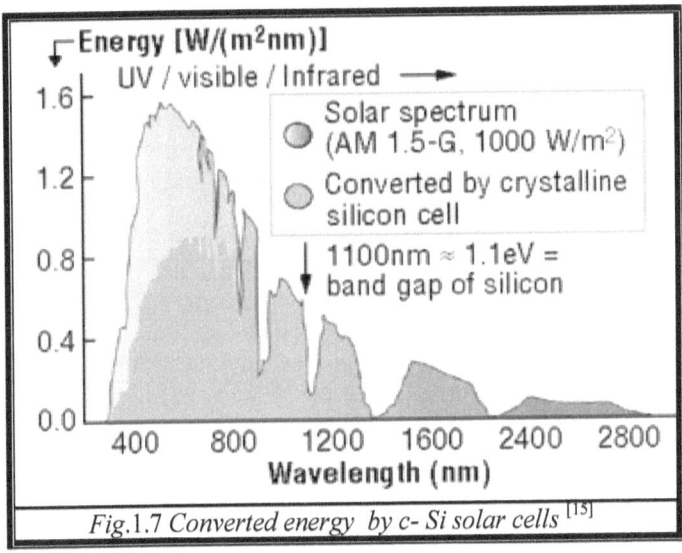

Fig.1.7 Converted energy by c-Si solar cells [15]

Recently, solutions to overcome the limitations of thin-film crystalline silicon have been developed [16]. Light trapping schemes where the weakly absorbed long wavelength light is obliquely coupled into the silicon and traverses the film several times can significantly enhance the absorption of sunlight in the thin silicon films. Minimizing the top contact coverage of the cell surface is another method for reducing optical losses; this approach simply aims at reducing the area that is covered over the cell to allow for maximum light input into the cell. Anti-reflective coatings can also be applied to create destructive interference within the cell. This can be done by modulating the refractive index of the surface coating; if destructive interference is achieved, there will be no reflective wave and thus all light will be transmitted into the semiconductor cell. Surface texturing is another option, but may be less viable because it also increases the manufacturing price. By applying a texture to the surface of the solar cell, the reflected light can be refracted into striking the surface again, thus reducing the overall light reflected out. Light trapping as another method allows for a decrease in overall thickness of the device; the path length that the light will travel is several times the actual device thickness. This can be achieved by adding a textured back reflector to the device as well as texturing the surface. If both front and rear surfaces of the device meet this criterion, the light will be 'trapped' by not having an immediate pathway out of the device due to internal reflections. As well, can be use solar concentrators such as microlens arrays or waveguides can be used to get a high optical efficiency due to increasing solar flux [16].

1.6 Acceptance angle

Acceptance angle is the maximum angle at which incoming sunlight can be captured by a solar concentrator. Its value depends on the concentration of the optic and the refractive index in which the receiver is immersed. Maximizing the acceptance angle of a concentrator is desirable in practical systems and it may be achieved by using non-imaging optics [17].

The simplest concentrator is a lens with a receiver R as shown in figure 1.8. The left section of the figure shows a set of parallel rays incident on the concentrator at an angle (α) less than acceptance angle (θ) to the optical axis. All rays end up on the receiver and, therefore, all light is captured. In the center, this figure shows another set of parallel rays, now incident on the concentrator at an angle $\alpha = \theta$. For an ideal concentrator, all rays are still captured. However, on the right, this figure shows yet another set of parallel rays, now incident on the concentrator at an angle $\alpha > \theta$. All rays now miss the receiver and all light is lost. Therefore, for incidence angles $\alpha < \theta$ all light is captured while for incidence angles $\alpha > \theta$ all light is lost. The concentrator is then said to have an acceptance angle θ or a total acceptance angle 2θ (since it accepts light within an angle $\pm\theta$ to the optical axis).

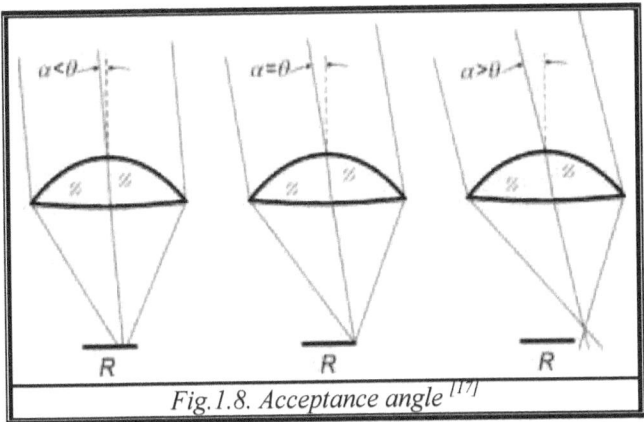

Fig.1.8. Acceptance angle [17]

Ideally, a solar concentrator has a transmission curve c_I as shown in figure 1.9. Transmission (efficiency) is $\tau = 1$ for all incidence angles $\alpha < \theta_I$ and $\tau = 0$ for all incidence angles $\alpha > \theta_I$. In practice, real transmission curves are not perfect and they typically have a shape similar to that of curve c_R, which is normalized so that $\tau = 1$ for $\alpha = 0$. In that case, the real acceptance angle θ_R is typically defined as the angle for which transmission τ drops to 90% of its maximum [17].

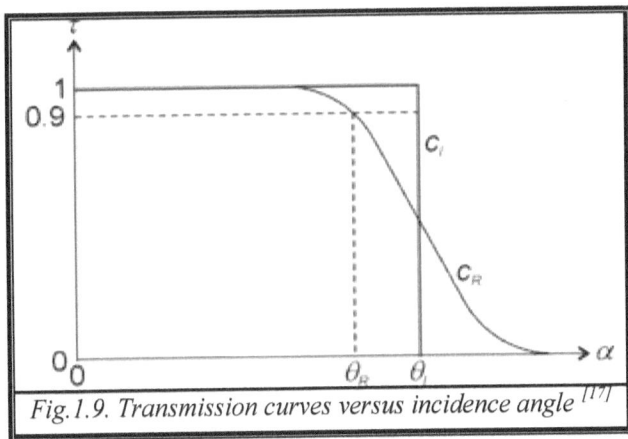

Fig.1.9. Transmission curves versus incidence angle [17]

The concentrator must still has enough acceptance angle to capture sunlight, also must has some angular dispersion θ_S when it is seen from the earth. It is, therefore, very important to design a concentrator with the widest possible acceptance angle. That is possible using non-imaging optics, which maximize the acceptance angle for a given concentration [17].

Sunlight is not a set of perfectly parallel rays (blue rays in figure 1.10), but it has a given angular aperture θ_S (equal to ±0.25°), as indicated by the (green rays). If the acceptance angle of the optic is wide enough, sunlight incident along the optical axis will be captured by the concentrator. However, for wider incidence angles α some light may be lost, as shown on the right. Perfectly parallel rays (blue rays) would be captured, but sunlight, due to its angular aperture, is partially lost.

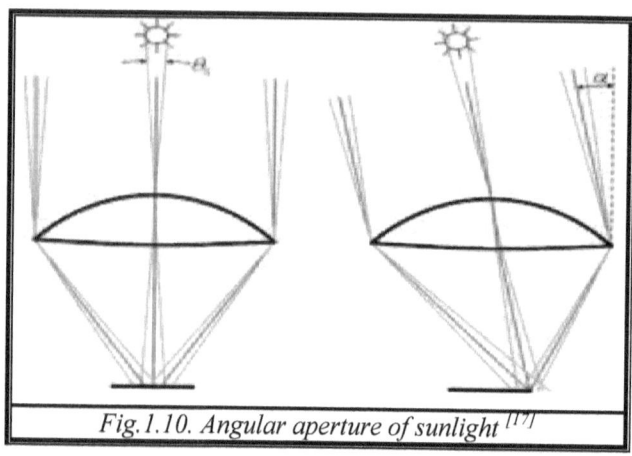

Fig.1.10. Angular aperture of sunlight [17]

Parallel rays and sunlight are therefore transmitted differently by a solar concentrator and the corresponding transmission curves are also different. Different acceptance angles may then be determined for parallel rays or for sunlight. For a given acceptance angle θ, the maximum concentration possible, C_{max} is given by [17]:

$$C_{max} = \frac{n^2}{sin^2\theta} \qquad (1.3)$$

Where n is the refractive index of the medium in which the receiver is immersed. In practice, real concentrators either have a lower than ideal concentration for a given acceptance or they have a lower than ideal acceptance angle for a given concentration. This can be summarized in the expression [17]:

$$CAP = \sqrt{C}sin\theta \leq n \qquad (1.4)$$

Which defines a quantity CAP (concentration acceptance product), which must be smaller than the refractive index of the medium in which the receiver is immersed. The higher CAP means the maximum concentration and acceptance angle.

Chapter two

Micro Optics

Chapter two

Micro Optics

2.1 Introduction

A microlens is a small lens, generally with a diameter less than a two millimeter[18], the small sizes of the lenses mean that a simple design can give good optical quality but sometimes unwanted effects arise due to optical diffraction at the small features. A typical microlens may be a single element with one plane surface and one spherical convex surface to refract the light. Because microlens are so small, the substrate that supports them is usually thicker than the lens and this has to be taken into account in the design. More sophisticated lenses may use aspherical surfaces and others may use several layers of optical material to achieve their design performance.

A different type of microlens has two flat and parallel surfaces and the focusing action is obtained by a variation of refractive index across the lens. These are known as gradient-index (GRIN) lenses. Some microlens achieve their focusing action by both a variation in refractive index and by the surface shape [18].

Another class of microlens, sometimes known as micro-Fresnel lens, focuses light by refraction in a set of concentric curved surfaces. Such lenses can be made very thin and lightweight. Binary-optic microlens focus light by diffraction. They have grooves with stepped edges or multilevel that approximates the ideal shape. They have advantages in fabrication and replication by using standard semiconductor processes such as photolithography and reactive ion etching (RIE) [19].

2.2 Microlens Arrays

Microlens arrays contain multiple lenses formed in a one-dimensional or two-dimensional array on a supporting substrate. If the individual lenses have circular apertures and are not allowed to overlap, they may be placed in a hexagonal array to obtain maximum coverage of the substrate. However there will still be gaps between the lenses which can only be reduced by making the microlens with non-circular apertures (rectangular or triangular apertures). With optical sensor arrays tiny lens systems serve to focus and concentrate the light onto the photodiode surface instead of allowing it to fall on non-photosensitive areas of the pixel device. Fill-factor is the ratio of the active refracting area, i.e. that area which directs light to the photosensor, to the total contiguous area occupied by the microlens arrays [20].

Similar with classical lenses, there are parameters like focal length, aperture, diameter, and surface shape which specify the microlens. But, microlens arrays require additional specifications on aperture geometry of the individual lenses, as well as on geometry and size of array. In principle, a wide range of aperture geometries can be realized. However, lens arrangement, application requirement, and the production technology chosen are leading typically to quadratic, rectangular, hexagonal or circular aperture (figure 2.1). Mostly, the lens shape is cylindrical or spherical. Array size strongly depends on the applications. It can range from 1×2 for optical switch applications up to more than $10^3 \times 10^3$ for diffusers or display applications [21].

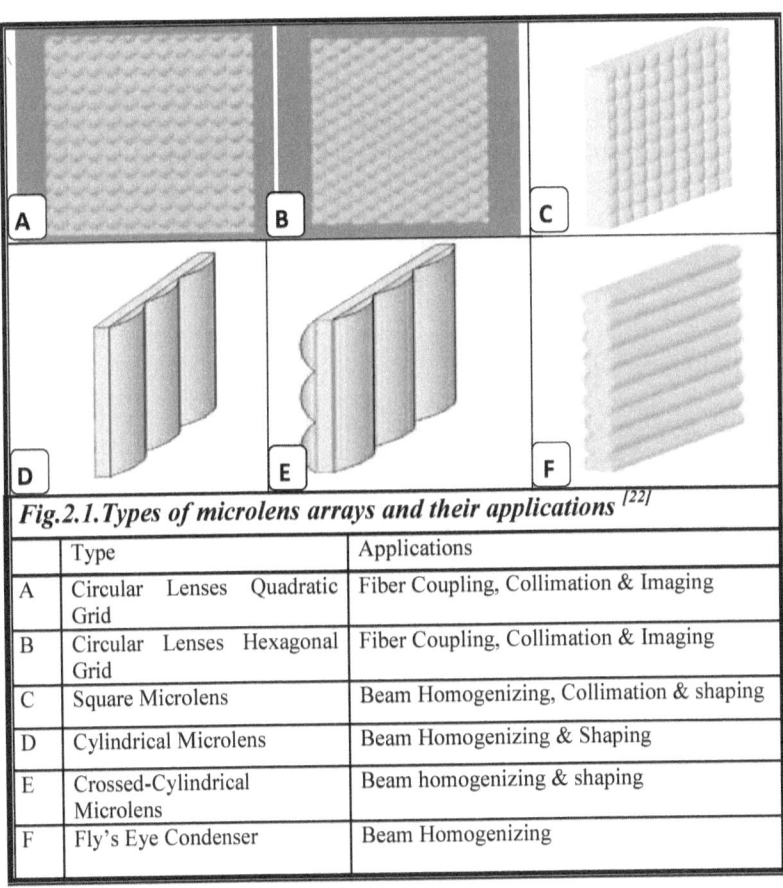

Fig.2.1.Types of microlens arrays and their applications [22]

	Type	Applications
A	Circular Lenses Quadratic Grid	Fiber Coupling, Collimation & Imaging
B	Circular Lenses Hexagonal Grid	Fiber Coupling, Collimation & Imaging
C	Square Microlens	Beam Homogenizing, Collimation & shaping
D	Cylindrical Microlens	Beam Homogenizing & Shaping
E	Crossed-Cylindrical Microlens	Beam homogenizing & shaping
F	Fly's Eye Condenser	Beam Homogenizing

Single microlens is used to couple light to optical fibers while microlens arrays are often used to increase the light collection efficiency of charge-coupled device (CCD) array. They collect and focus light that would have otherwise fallen on to the non-sensitive areas of the CCD. Microlens arrays are also used in some digital projectors, to focus light to the active areas of the LCD used to generate the image to be projected. Current research also relies on microlens of various types to act as concentrators for high efficiency photovoltaic for electricity production [22].

Combinations of microlens arrays have been designed so that they have novel imaging properties, such as the ability to form an image at unit magnification and not inverted (fly eye optical system) as is the case with conventional lenses. Microlens arrays have been developed to form compact imaging devices for applications such as photocopiers and mobile-phone cameras. Another application is in 3D imaging and displays [23].

2.3 Micro Optics in Nature

Examples of micro optics are to be found in nature ranging from simple structures to gather light for photosynthesis in leafs to compound eyes in insects, as methods of forming microlens and detector arrays are further developed then the ability to mimic optical designs found in nature will lead to new compact optical systems[24]. For all types of creatures evolution has found appropriate image capturing systems to give them all necessary visual information about their environment (figure 2.2), and the types of natural eyes is:

2.3.1 Single-aperture eyes

For large vertebrates like humans, eyes are optimized to provide a high resolution, large field of view, focusing ability, color detection and a very large dynamic range to see both in the bright sunshine and in the dark night. Here, the size or volume of the eye is a free parameter for the design. The optical performance is the key issue. Single-aperture eyes (like the human eyes) are similar to photographic or electronic camera systems. The eye consists of a flexible lens for focusing, a variable pupil (iris) for fast sensitivity adaptation and the retina, the image detector. The field of vision of a human eye approximates an ellipse about 150° high by about 210° wide. The angular resolution or acuity is around (0.6 – 1) min of arc for the fovea. A large single-pupil eye is a perfect solution if miniaturization is not an issue [24].

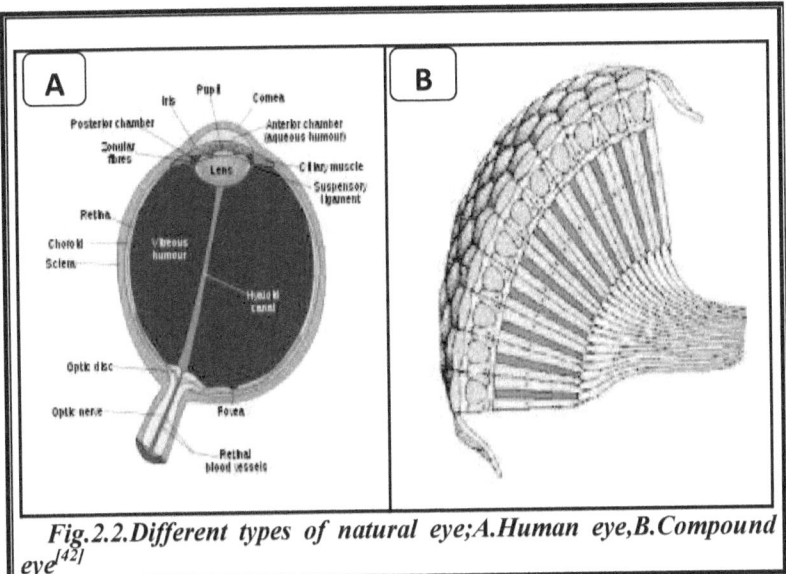

Fig.2.2.Different types of natural eye;A.Human eye,B.Compound eye[42]

2.3.2 Compound eyes

For small invertebrates having an external skeleton, eyes are very expensive in weight and metabolic energy consume. If the budget is tight, nature prefers to distribute the image capturing to a matrix of some small eye sensors instead of using a single eye. The resolution of such so-called compound or fly's eyes is usually poor compared to the single-aperture eyes. In nature, this lack of resolution is often counterbalanced by additional functionality like a very large view angle, polarization or fast movement detection. Natural eyes are a perfect compromise suited to the requirements of the lifestyle of the animal [25].

Compound eyes are multi-aperture optical sensors of insects and crustaceans and generally are divided into two main classes: apposition compound eyes and superposition compound eyes (figure 2.3):

2.3.2.1 Apposition and superposition

Apposition eye consists of an array of lenses and photoreceptors each of the lenses focusing light from a small solid angle of object space onto a single photoreceptor. Each lens-photoreceptor system is referred to as ommatidia (the imaging units of insects). Apposition eyes have some hundreds up to tens of thousands of these ommatidia packed in non-uniform hexagonal arrays.

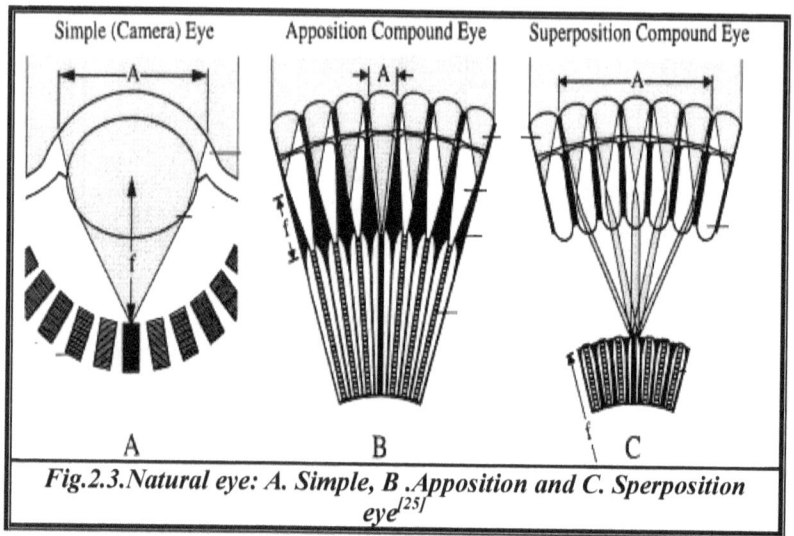

Fig.2.3.Natural eye: A. Simple, B .Apposition and C. Sperposition eye[25]

Superposition compound eye has primarily evolved on nocturnal insects and deep-water crustaceans. The light from multiple facets combines on the surface of the photoreceptor layer to form a single erect image of the object. Compared to apposition eyes, the superposition eye is much lighter sensitive. Some insects use a combination of both types of compound eyes. Switch between apposition (daylight) and superposition (night) or change the number of recruited facets making up the superposition image. A very interesting modification is the neural superposition eye of a housefly, where each ommatidia has seven detector pixels [25].

Signals of different adjacent ommatidia interact within the neurons of the fly's brain system. This allows fast directionally selective motion detection, which enables the housefly to maneuver perfectly in three dimensional spaces.

2.4 Nano-Optics

Nano-optics is the study of the behavior of light on the nanometer scale. It is considered as a branch of optical engineering which deals with optics, or the interaction of light with particles or substances, at deeply subwavelength length scales [26]. Technologies in the realm of nano-optics include near-field scanning optical microscopy (NSOM), photo-assisted scanning tunneling microscopy, and surface Plasmon optics. Traditional microscopy makes use of diffractive elements to focus light tightly in order to increase resolution. But because of the diffraction limit (also known as the Rayleigh Criterion); propagating light may be focused to a spot with a minimum diameter of roughly half the wavelength of the light. Thus, even with diffraction-limited confocal microscopy, the maximum resolution obtainable is on the order of a couple of hundred nanometers. The scientific and industrial communities are becoming more interested in the characterization of materials and phenomena on the scale of a few nanometers, so alternative techniques must be utilized. Scanning Probe Microscopy (SPM) makes use of a "probe", (usually either a tiny aperture or super-sharp tip), which either locally excites a sample or transmits local information from a sample to be collected and analyzed. The ability to fabricate devices in nano-scale that has been developed recently provided the catalyst for this area of study. The study of nano-photonics involves two broad themes: studying the novel properties of light at the nanometer scale, enabling highly power efficient devices for engineering applications. The study has the potential to revolutionize the telecommunications industry by providing low power, high speed, interference-free devices such as electro optics and all-optical switches on a chip [26].

2.5 Optical Systems

Optical systems cover a wide range of specifications which depend on the design requirements. These specifications serve as the goal for the design and construction of the optical system. In addition, these specifications are bases for tolerances placed upon the components of the optical system and lead to detailed component specifications used for procurement of the optical elements of the system. Optical systems can classify to consider of functional specifications into imaging and non-imaging optical systems [27].

2.5.1 Imaging Optical System

Imaging optical system is a system capable of being used for imaging. The two traditional systems are mirror systems and lens systems, although in the late twentieth century, optical fiber was introduced. Mirrors and lenses have a focal point, while optical fiber transfers an image from one plane to another without an optical focus. In physical optics, light is considered to propagate as a wave. This model predicts phenomena such as diffraction and aberration:

2.5.1.1 Diffraction

Diffraction occurs in all imaging optical systems alike, and is defined as the phenomenon of deviation of the beam of light on its path when it passed through a narrow aperture or a sharp edge. Diffraction is a natural property of light arising from its wave nature, possesses fundamental limitation on any optical system. Diffraction is always present, although its effects may be made if the system has significant aberrations. When an optical system is essentially free from aberrations, its performance is limited solely by diffraction, and it is referred to as diffraction-limited [28].

Diffraction arises because of the way in which waves propagate; this is described by the Huygens principle and the principle of superposition of waves. The wave displacement at any subsequent point is the sum of these secondary waves. When waves are added together, their sum is determined by the relative phases as well as the amplitudes of the individual waves so that the summed amplitude of the waves can have any value between zero and the sum of the individual amplitudes. Hence, diffraction patterns usually have a series of maxima and minima [28].

2.5.1.2 Aberration

Aberration is a departure of the performance of an optical system from the predictions of paraxial optics. In an imaging system, it occurs when light from one point of an object does not converge into (or does not diverge from) a single point after transmission through the system. Aberrations occur because the simple paraxial theory is not a completely accurate model of the effect of an optical system on light, rather than due to flaws in the optical elements[29].

Aberration leads to blurring of the image produced by an image-forming optical system. Makers of optical instruments need to correct optical systems to compensate for aberration.

Aberrations fall into two classes: monochromatic and chromatic. Monochromatic aberrations are caused by the geometry of the lens or the mirror and occur both when light is reflected and when it is refracted. They appear even when using monochromatic light; it is classified into five types: Spherical Aberration, Coma Aberration, Astigmatism, Field Curvature and Distortion [30].

Chromatic aberrations are caused by dispersion, the variation of a lens's refractive index with wavelength, so they do not appear in mirrors, and do not appear when monochromatic light is used. It is classified into two types: longitudinal and transverse (figure 2.4).

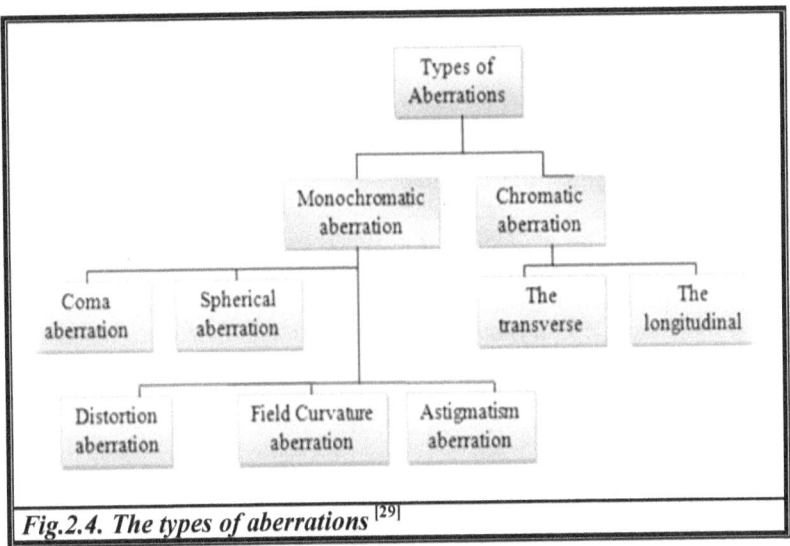

Fig.2.4. The types of aberrations [29]

2.5.2 Nonimaging Optical System

Nonimaging optics is the branch of optics concerned with the optimal transfer of light radiation between a source and a target. Unlike traditional imaging optics, the techniques involved do not attempt to form an image of the source; instead an optimized optical system for optical radiative transfer from a source to a target is desired [31].

The two design problems that nonimaging optics solves better than imaging optics are [32]:

- Solar energy concentration: maximizing the amount of energy applied to a receiver, typically a solar cell or a thermal receiver
- Illumination: controlling the distribution of light, typically so it is "evenly" spread over some areas and completely blocked from other areas

Typical variables to be optimized at the target include the total radiant flux, the angular and spatial distribution of optical radiation. These variables on the target side of the optical system often must be optimized while simultaneously considering the collection efficiency of the optical system at the source.

2.5.2.1 Characteristics of nonimaging optical system

For a given concentration, nonimaging optics provide the widest possible acceptance angles and, therefore, are the most appropriate for use in solar concentration as, for example, in concentrated photovoltaics. When compared to

"traditional" imaging optics (such as parabolic reflectors or Fresnel lenses), the main advantages of nonimaging optics for concentrating solar energy are [33]:
1. wider acceptance angles for:
   - less precise tracking
   - imperfectly manufactured optics
   - imperfectly assembled components
   - movements of the system due to wind
   - finite stiffness of the supporting structure
   - deformation due to aging
   - capture of circumsolar radiation
   - other imperfections in the system
2. higher solar concentrations for:
   - smaller solar cells (in concentrated photovoltaics)
   - higher temperatures (in concentrated solar thermal)
   - lower thermal losses (in concentrated solar thermal)
   - widen the applications of concentrated solar power, for example to solar lasers
3. possibility of a uniform illumination of the receiver for:
   - improve reliability and efficiency of the solar cells (in concentrated photovoltaics)
   - improve heat transfer (in concentrated solar thermal)
4. design flexibility for: different kinds of optics with different geometries can be tailored for different applications

Also, for low concentrations, the very wide acceptance angles of nonimaging optics can avoid solar tracking altogether or limit it to a few positions a year.

The main disadvantage of nonimaging optics when compared to parabolic reflectors or Fresnel lenses is that, for high concentrations, they typically have one more optical surface, slightly decreasing efficiency. That, however, is only noticeable when the optics is aiming perfectly towards the sun, which is typically not the case because of imperfections in practical systems [33].

2.6 Computer Performance Evaluation

The performance characteristics of an imaging and nonimaging optical system can be represented in many ways. Often the final optical performance specification is in terms of the modulation transfer function (MTF), encircled energy, RMS blur diameter (spot diagram), or other criteria. These criteria relate in different ways to the image quality of the system. Image quality can be thought of as resolution or how close two objects can approach each other while still being resolved or distinguished from one another. Image quality can also be thought of as image sharpness, crispness, or contrast [34].

Image is never perfect. It is limited by geometrical aberrations, diffraction, the effects of manufacturing and assembly errors, and other factors. The characterization of image quality by the methods described in the following sections will help to assess just how system performs with respect to its image.

It is important to realize that the image quality or resolution of the entire system is not totally dependent on the optics, but may include the sensor, electronics, display device, and/or other system components making up the system. For example, if the eye is the sensor, it can accommodate for both defocus and field curvature, whereas a flat sensor such as a CCD cannot [34].

Worth mentioning, nonimaging optical system have good performance evaluation by encircled energy and spot diagram only, in place of other methods which evaluate image performance in imaging optical system.

2.6.1 Optical Resolution

Optical resolution describes the ability of an imaging system to resolve detail in the object that is being imaged. An imaging system may have many individual components including a lens and recording and display components. Each of these contributes to the optical resolution of the system, as will the environment in which the imaging is done[35].

The resolution of a system is based on the minimum distance at which the points can be distinguished as individuals. Several standards are used to determine, quantitatively, whether or not the points can be distinguished. One of the methods specify that, on the line between the center of one point and the next, the contrast between the maximum and minimum intensity is at least 26% lower than the maximum as shown in figure 2.5. This corresponds to the overlap of one Airy disk on the first dark ring in the other. This standard for separation is also known as the Rayleigh criterion in symbols, the distance is defined as follows:

$$d = 1.22\lambda \ f/\# \qquad (2.1)$$

Where d is the minimum distance between resolvable points, in the same units as wavelength λ is specified, and $f/\#$ is the focal number of the optical system.

2.6.2 Transverse Ray Aberration Curves

Transverse ray aberration curves or ray trace curves can immediately tell just how much spherical aberration, coma, astigmatism, field curvature, axial color, lateral color, and field curvature are present. In addition, in many cases the user can also tell what orders of these aberrations are present. Also, can often makes a reliable judgment to what to do next regarding further optimization of the lens. In spite of some fabulous advances in performance simulation and modeling, transverse ray aberration curves are still invaluable to the serious designer (figure 26) [36].

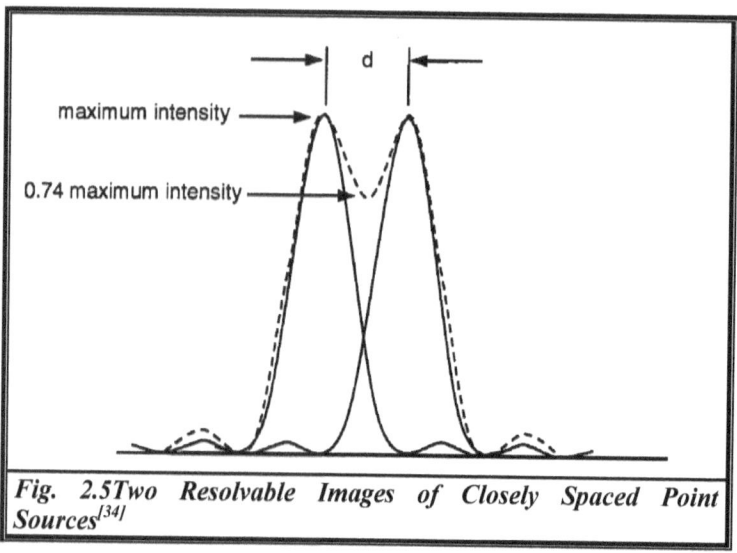

Fig. 2.5 Two Resolvable Images of Closely Spaced Point Sources[34]

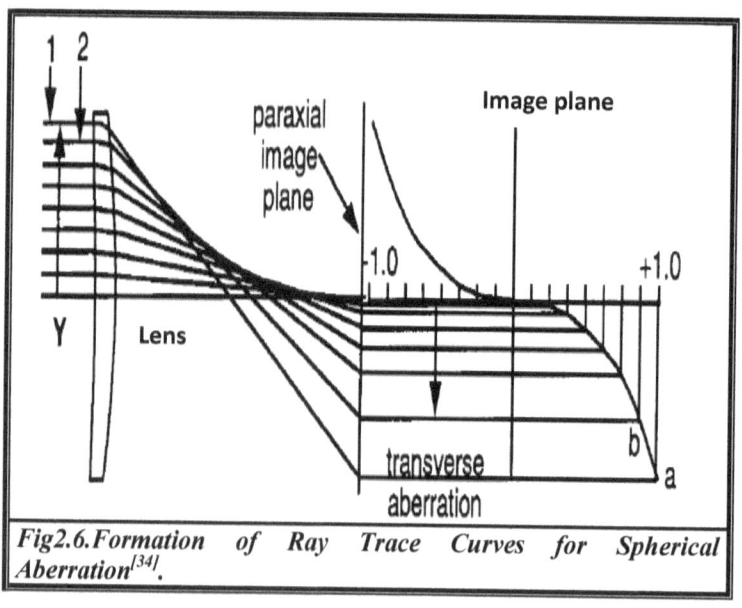

Fig 2.6. Formation of Ray Trace Curves for Spherical Aberration[34].

2.6.3 Optical Path Difference (OPD)

Optical path difference is an extremely useful measure of the performance of an imaging optical system. If the wavefronts proceeding to a given point image are spherical, concentric, and centered at the point image for a given field of view, then the imagery will be geometrically perfect, or diffraction limited [36].

Rays are perpendicular to the wavefronts. It is thus clear that if the wavefronts are spherical, concentric, and centered at a point in the image, then the rays will all come to that same point as defined by the center of curvature of the wavefronts.

Figure 2.7 shows a hypothetical lens with a perfectly spherical reference wavefront and an aberrated wavefront. The aberrated wavefront departs from sphericity due to aberrations induced by the lens. OPD is the difference between the real wavefront and a spherical reference wavefront, which is usually selected to be a near best fit to the aberrated wavefront.

One of the reasons the OPD is so valuable a parameter is evident from the Rayleigh criteria; showed that an optical instrument would not fall seriously short of the performance possible with an absolutely perfect system if the distance between the longest and shortest paths leading to a selected focus did not exceed one-quarter of a wavelength.

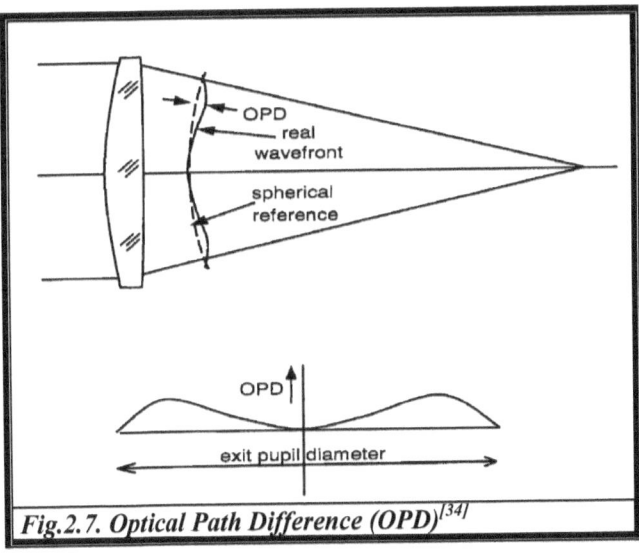

Fig.2.7. Optical Path Difference (OPD) [34]

2.6.4 Optical Transfer Function (OTF)

The optical transfer function (OTF) of an imaging system (camera, video system, microscope etc.) is the true measure of resolution (image sharpness) that the system is capable of. The common practice of defining resolution in terms of pixel count is not meaningful, as it is the overall OTF of the complete system, including lens as well as other factors, that defines true performance. The optical transfer function is roughly the equivalent of phase and frequency response in an audio system, and can be represented by a graph of light amplitude (brightness) and phase versus spatial frequency (cycles per picture width) [36].

In the most common applications (cameras and video systems) it is the modulation transfer function (MTF) (figure 2.8), the magnitude component of the OTF that is most relevant, although the phase component can have a secondary effect. While optical resolution, as commonly used with reference to camera systems, describes only the number of pixels in an image, and hence the potential to show fine detail, the transfer function describes the ability of adjacent pixels to change from black to white in response to patterns of varying spatial frequency, and hence the actual capability to show fine detail, whether with full or reduced contrast [36].

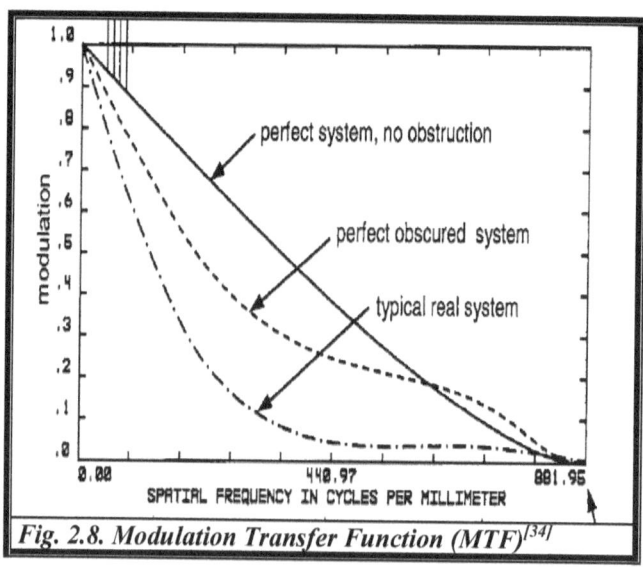

Fig. 2.8. Modulation Transfer Function (MTF)[34]

2.6.5 Spot Diagrams

Spot diagrams are the geometrical image blur formed by the lens when imaging a point object. This is more functionally useful form of output. The spot diagrams give a visual representation of the energy distribution in the image of a point object (figure 2.9). Generally, the RMS spot radius is output with the spot diagrams. The RMS spot diameter is the diameter of a circle containing approximately 68% of the energy. This metric can be of great value, especially when working with pixelated sensors where one often wants the image of a point object to fall within a pixel [37].

The spot diagram data may be used to obtain useful information regarding the quality of the image, as the geometrical spot size of the image and the radial energy distribution. The image of a point by a lens that is not diffraction limited is often described by its geometrical spot size, defined to be the RMS spot radius (not diameter). Although this quantity does not indicate the fractional energy in the spot. Calculating the spot size accurately is thus a matter of considerable importance in optical design software [37].

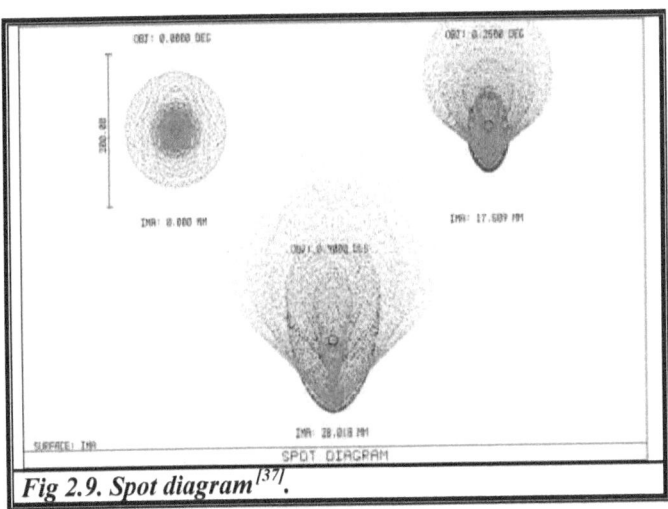

Fig 2.9. Spot diagram[37].

2.6.6 Encircled Energy

The optics term encircled energy refers to a measure of concentration of energy in an optical image, or projected laser at a given range. If a single point source is brought to its sharpest focus by a lens giving the smallest image possible with that given lens (called a point spread function or PSF), calculation of the encircled energy of the resulting image gives the distribution of energy in that PSF [38].

Encircled energy is calculated by first determining the total energy of the PSF over the full image plane, then determining the centroid of the PSF. Circles of increasing radius are then created at that centroid and the PSF energy within each circle is calculated and divided by the total energy. As the circle increases in radius, more of the PSF energy is enclosed, until the circle is sufficiently large to completely contain all the PSF energy. The encircled energy curve thus ranges from zero to one as shown in figure 2.10 [38].

A typical criterion for encircled energy is the radius of the PSF at which either 50% or 80% of the energy is encircled. This is a linear dimension, typically in micrometers. When divided by the lens or mirror focal length, this gives the angular size of the PSF, typically expressed in arc-seconds when specifying astronomical optical system performance.

Encircled energy is also used to quantify the spreading of a laser beam at a given distance. All laser beams spread due to the necessarily limited aperture of the optical system projecting the beam. As in star image PSF's, the linear spreading of the beam expressed as encircled energy is divided by the projection distance to give the angular spreading.

An alternative to encircled energy is ensquared energy, typically used when quantifying image sharpness for digital imaging cameras using pixels [38]. Ensquared energy has been used in this work as a replacement of encircled energy to cover all square area of PVC.

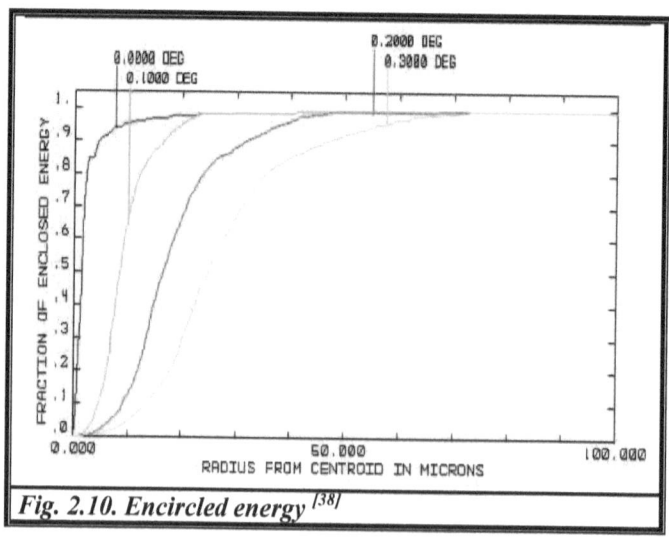

Fig. 2.10. Encircled energy [38]

Chapter three

Optical Design

Chapter three

Optical Design

3.1 Introduction

Optical design refers to the calculation of optical construction parameters (variables) that will meet a set of performance requirements and constraints, including cost and manufacturing limitations. Construction parameters include surface profile types (spherical, aspheric, holographic, diffractive, etc.), and the parameters for each surface type such as radius of curvature, distance to the next surface, glass type and optionally tilt and decenter. Performance requirements include [38]:

1. Optical performance, i.e., image quality: quantified by encircled energy, MTF, Strehl ratio, reflection control, and pupil performance (size, location and aberration control); the choice of the image quality metric is application specific.
2. Physical requirements such as weight, static volume, dynamic volume, center of gravity and overall configuration requirements.
3. Environmental requirements: ranges for temperature, pressure, vibration and electromagnetic shielding.

Design constraints can include realistic lens element center and edge thicknesses, minimum and maximum air-spaces between lenses, maximum constraints on entrance and exit angles, physically realizable glass index of refraction and dispersion properties.

Manufacturing costs and delivery schedules are also a major part of optical design. The price of an optical glass blank of given dimensions can vary, depending on the size, glass type, index homogeneity quality, and availability, with borosilicate glass BK7 usually being the cheapest. This is primarily due to increased blank annealing time required to achieve acceptable index homogeneity. Availability of glass blanks is driven by how frequently a particular glass type is mixed and poured by a given manufacturer, and can seriously affect manufacturing cost and schedule [40].

3.2 Optical Software Programs

Modern optical software programs provide engineers and designers with fast, accurate virtual-prototyping resources to visualize optical system designs. The right software package improves engineer productivity and reduces product development costs, leading to better innovations on tighter budgets and timelines. There are a number of options available to choice these programs that fit into three categories: sequential ray-tracing; non-sequential ray-tracing; and finite-difference time-domain (FDTD) simulation[41]. The most important programs are summarized in table 3.1.

Table 3.1. Optical Software Programs[87]	
Categories	Programs
sequential ray-tracing programs	Code V, OpTaliX, OSLO and ZEMAX
non-sequential ray-tracing programs	ASAP, FRED, Light Tool, SPEOS, Trace Pro and ZEMAX
FDTD programs	FDTD Solutions, Full Wave, JCM suite, Omnisim and Opti FDTD

3.3 ZEMAX Software Program

ZEMAX is a widely-used optical design program [42]. It is used for the design and analysis of optical systems. ZEMAX can perform standard sequential ray tracing through optical elements, non-sequential ray tracing for analysis of stray light, and physical optics beam propagation.

ZEMAX is used for the design of optical systems such as camera lenses and analysis of illumination systems. It can model the propagation of rays through optical elements such as lenses (including aspheres and gradient index lenses), mirrors, and diffractive optical elements. ZEMAX can model the effect of optical coatings on the surfaces of components, and can produce standard analysis diagrams such as spot diagrams and ray-fan plots. It includes an extensive library of stock lenses from a variety of manufacturers. The physical optics propagation feature can be used for problems where diffraction is important, including the propagation of laser beams, holography, and the coupling of light into single-mode optical fibers [43].

3.3.1 Optimization

ZEMAX has a powerful suite of optimization tools that can be used to optimize a lens design by automatically adjusting parameters to maximize performance and reduce aberrations [90]. It also has an extensive tolerancing capability. The processes of optimization are very powerful and able to improve the designs of the optical system and this requires an acceptable starting point and group of parameters. Any measure describes the lens can be a variable. Normally it is used only one parameter in order to judge the control on some properties of the lens. The most important parameters that conducting on them the process of optimization are [44].

1. Radius of curvature of the surfaces.
2. Space between optical surfaces which can be the thickness of the element or air space between the elements.
3. Type of optical glass used to manufacture the lens.

3.3.2 Merit Function

Merit function is the function used by a perfect operation of program (optimization) , which we will have in the design steps, to improve the parameters and thus get closer to the desired design. So that the design parameters can be included such as weight , size , target , coefficients of aberration and ray because it consists of a group of deviations and aberrations that can be corrected or minimized to the limited value. This function is evaluated in each stage of the design to determine the design conformity in current step with required design. The reduce value of the function must be interpreted that the design is closer to the desired result .The Merit function is the numerical representation to how the conformity of the optical system largely to specific group of the goals design requirements [44].

The ideal case is the required condition of the design and it use a wide range of parameters to represent the controls or different aims of the system and be specific and required from the design such as (effective focal length, F/# (f-number), the site of the aperture, the purity of the image, type of glass used to manufacture of the lens , elements diameters of components of the system) [45] .

3.4 Ray-Tracing Mode

The general ray tracing mode consists of sources, optical path, and receivers. Figure 3.1 shows the drawing of the general ray tracing mode including the property definitions. The source defines the illumination conditions presented to the optical system, specifically spectrum, divergence, source size, and source position. The receiver records incident rays at the target plane. The receiver provides a number of metrics including total incident power, irradiance distributions, intensity (angular) distributions, and spectral information. The optical path is defined by elements, each of which has surface, material, and interface properties [92].

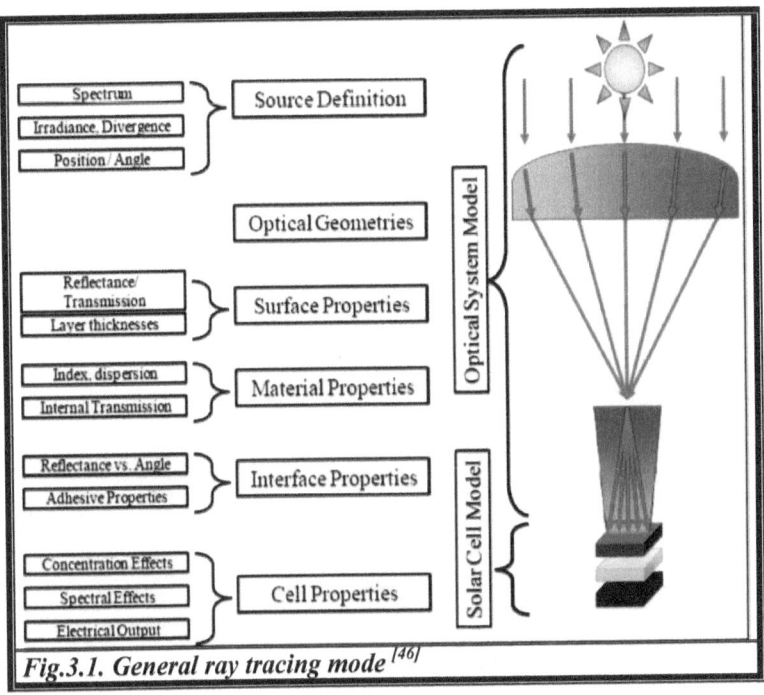

Fig.3.1. General ray tracing mode [46]

3.5 Sequential Ray-Tracing Mode in ZEMAX

Ray tracing is a widely applicable technique for modeling the propagation of light through an optical system. The modeling of light propagation via ray tracing is commonly called geometrical optics.

In sequential ray tracing, rays are traced through a predefined sequence of surfaces while traveling from the object surface to the image surface. Rays hit each surface once in the order (sequence) in which the surfaces are defined. Imaging systems are well described by sequential surfaces. Sequential ray tracing is numerically fast and is extremely useful for the design, optimization and tolerancing of such systems. Ray fan plots, MTF, OPD, spot diagram and encircled energy can be easily performed using sequential ray tracing [47].

Figure 3.2.showes the layout of ZEMAX sequential components. Includes the main menu bar, button bar and Lens Data Editor (LDE). All of the features that ZEMAX has to offer can be accessed through the various menus in the main menu bar. Shortcuts to most of these features are available for convenience in the button bar below the main menu bar. The assigned buttons can be changed via the menu option [47].

Beneath the button bar is the LDE. The LDE has columns for comments, radius, thickness, glass, and semi-diameter (radial clear aperture) and conic constant. The latter five data items are used to define the majority of optical components.

Each row corresponds to an optical surface. Each surface has its own local coordinate system. The position of each surface along the optical axis is referenced to the previous surface. In other words, the "thickness" column in the lens data editor refers to the distance from the previous surface and not from a global reference point [47].

By default, there are three surfaces shown: the object, stop, and image. These are denoted by OBJ, STO, and IMA in the small column on the left hand side. The second column also displays a surface type, the default of which is the "standard" surface. There are many other surface types available. The columns to the right of the conic constant column are used for setting additional parameters for more advanced surface types [47].

Every optical system has a system aperture specification, such as focal number F/#, entrance pupil diameter, numerical aperture NA, or cone angle. This sets the width of the on-axis beam that the optical system will collect in object space. In ZEMAX, this data is specified in the aperture tab of the general dialog. This can be accessed from the main menu via the menu option, "System → General". The field data dialog is used to specify the points on the object surface from which rays are launched. This dialog can be accessed from the "System →Fields" option in the main menu. The wavelengths of rays that are traced are set in the wavelength data dialog. This dialog is accessed from the main menu option [47].

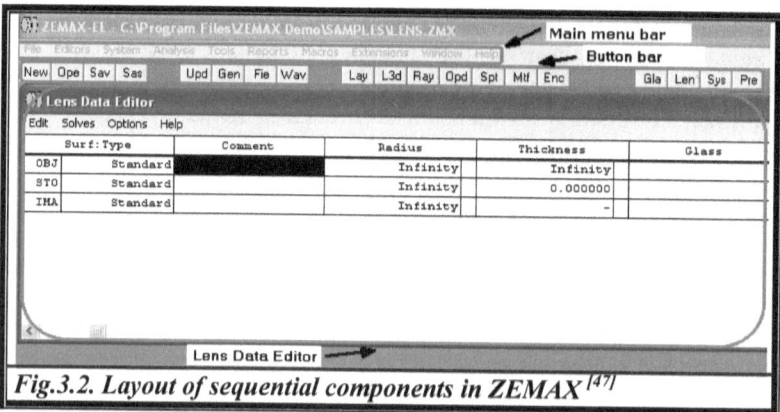

Fig.3.2. Layout of sequential components in ZEMAX [47]

3.6 Non-Sequential Ray-Tracing Mode in ZEMAX

Non-sequential ray tracing implies that there is no predefined sequence of surfaces which rays that are being traced must hit. The objects that the rays hit are determined solely by the physical positions and properties of the objects as well as the directions of the rays. Rays may hit any part of any non-sequential object, and may hit the same object multiple times, or not at all. This can be contrasted with sequential ray tracing where all of the rays traced must propagate through the same set of surfaces in the same order [47].

In sequential mode, all ray propagation occurs through surfaces which are located using a local coordinate system. In non-sequential mode, optical components are modeled as true three-dimensional objects, either as surfaces or solid volumes. Each object is placed globally at an independent (x, y, z) coordinate with an independently defined orientation [47].

The non-sequential ray tracing capabilities of ZEMAX do not suffer from the same limitations that sequential ray tracing does. Since rays can propagate through the optical components in any order, total internal reflection (TIR) ray paths can be accounted for. While sequential mode is limited to the analysis of imaging systems, non-sequential mode can be used to analyze stray light, scattering and illumination in both imaging and non-imaging systems. If an optical system can be traced with rays, it can be traced with non-sequential analysis in ZEMAX [47].

There are many types of optical components which cannot be modeled using the simple sequential surface model. Such optics need to be modeled as real, 3D components. Examples of objects that require non-sequential ray tracing include: complex prisms, corner cubes, light pipes, faceted objects and embedded volume objects (i.e. objects located within other objects) [47].

Non-sequential ray tracing can be modeled in ZEMAX using one of two modes: Pure non-sequential ray tracing and mixed sequential/non sequential ray tracing.

Rays from non-sequential sources, can be split and scattered by optical components. These rays can also be diffracted at phase surfaces/objects. The analysis options available when tracing non-sequential rays include evaluating radiometric data on detectors and the storing of ray data in ray database files. Detectors can be modeled as planar surfaces, curved surfaces and even three-dimensional volumes. Non-sequential detectors support the display of a variety of data types including: incoherent irradiance, coherent irradiance, coherent phase, radiant intensity and radiance. Ray database files store the history of each ray traced. Ray paths can be filtered to isolate rays that hit specific objects. The filtered ray data can then be displayed in layouts and on detector objects. All of the above make pure non-sequential ray tracing very useful for ghost analysis, stray light analysis as well as a variety of illumination applications (figure 3.3) [47].

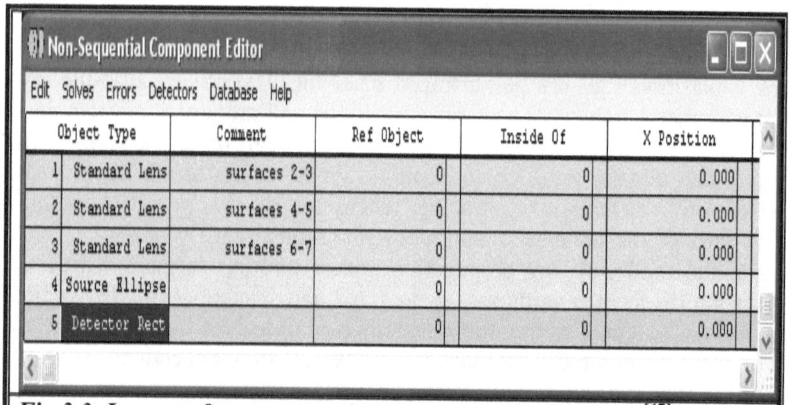

Fig.3.3. Layout of non-sequential components in ZEMAX [47]

3.7 Optical System prototypes that Used in ZEMAX

In this work, many optical system prototypes have been designed as a compact solar concentrator systems to improve optical efficiency. Some of these are functionally appropriate to use sequential ray tracing mode, like one and two dimensions Plano-convex microlens array, where sunlight collected by each lens of the array to focus into small area of PVC (figure 3.4).

Other prototypes are fit to work in non- sequential ray tracing mode; these prototypes composed of two dimensions Plano-convex microlens array, low refractive index cladding material and slab waveguide (figure 3.5), where sunlight collected by the array focuses onto localized prisms (facets) mirrors positioned in the bottom of slab to reflect light at angles that exceed the critical angle defined by Snell's law propagate via total internal reflection (TIR) within the waveguide to the exit aperture where PVC is fixed. There are two samples: double sides ray propagation sample (DRPS) and single side ray propagation sample (SRPS) into slab waveguide by controlling of prism's angle.

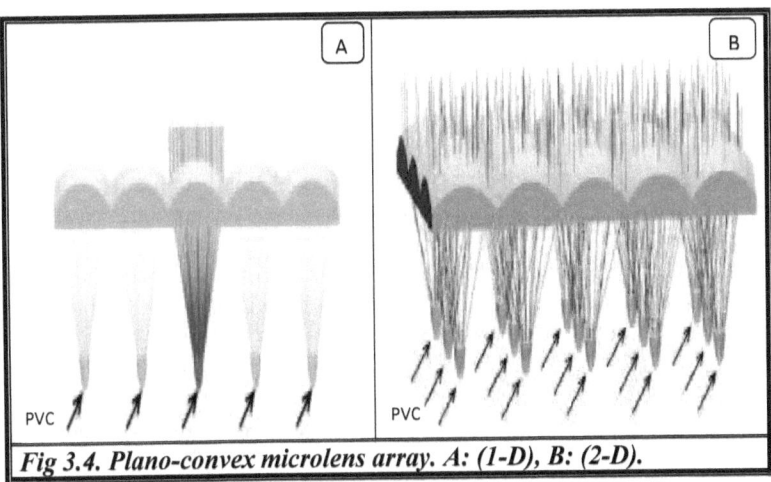

Fig 3.4. Plano-convex microlens array. A: (1-D), B: (2-D).

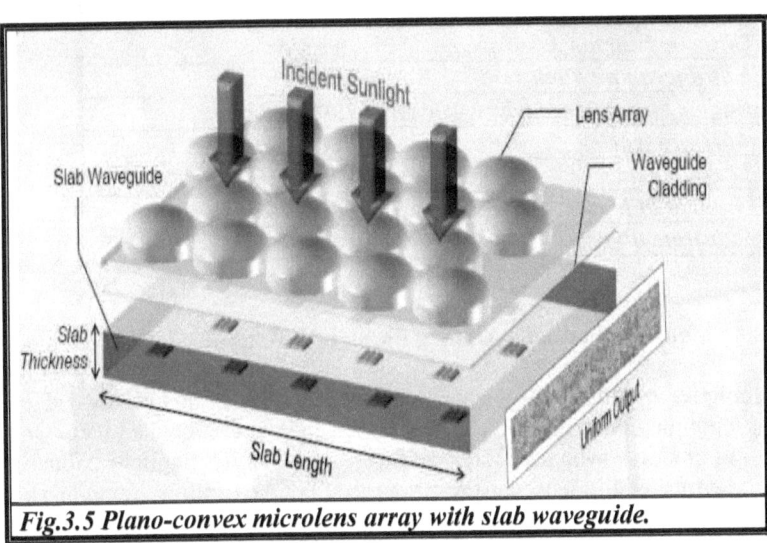

Fig.3.5 Plano-convex microlens array with slab waveguide.

3.7.1 Plano-Convex Microlens Array (1-D)

A one dimension microlens is characteristically composed of one plane surface and one spherical surface that are used to refract light. Microlens arrays are composed of several lenses that form a one-dimensional array on a supporting substrate. The design divides the upward-facing primary into several small apertures, each with its own individual secondary element and solar cell as shown in figure 3.4.A.

3.7.2 Plano-convex Microlens Array (2-D)

A two dimensions microlens is characteristically composed of one plane surface and one spherical surface that are used to refract light. Microlens arrays are composed of several lenses that form a two dimensional array on a supporting substrate. The design divides the upward-facing primary into several small apertures, each with its own individual secondary element and solar cell as shown in figure (3.4.B).

The choice of the optimum parameters of a microlens array depends very much on the actual design and requirements of optical system. The most parameters that are used in these samples summarized in table 3.2.

Table 3.2. parameters of 1-D & 2-D microlens array	
Parameter	Characteristics
Source	Sunlight
Glass (lens array & substrate)	Acrylic
Entrance pupil diameter	20 mm
Radius of curvature (convex lens)	1.041538 mm
Aperture type	Rectangular
Lens diameter	1 mm
Number of lenses	20x1 & 20x20
Substrate thickness	1 mm

3.7.3 Plano-Convex Microlens Array with Slab Waveguide (2-D)

An alternative approach for planar concentration has been investigated by replacing multiple nonimaging secondary optics and their associated PVC with a single multimode waveguide connected to a shared PVC. Sunlight collected by each aperture of the lenses array primary is coupled into a common slab waveguide using localized mirror prisms embedded on the backside of the waveguide to reorient focused light into guided modes that exceed the critical angle propagate via total internal reflection TIR within the waveguide to the exit aperture, typically at the edge of the slab. TIR is a complete reflection with negligible spectral or polarization-dependent losses which enables long propagation lifetimes. Guided rays can strike a subsequent coupling region and decouple as loss. The number of TIR interactions during propagation to the PVC affects the likelihood of decoupling and therefore the optical efficiency. Couplers typically cover (30% or 35%) of the waveguide surface enabling the system to yield both high efficiency and high concentration. The waveguide transports sunlight collected over the entire input aperture to a single PVC placed at the waveguide edge. PVC alignment becomes trivial since comparatively large cells are cemented to the waveguide edges. Fewer PVC's reduce connection complexity and allow one heat sink to manage the entire system output (figure 3.5).

The choice of the optimum parameters of a microlens array depends very much on the actual design and requirements of optical system. The most parameters that are used in this sample summarized in table 3.3.

Table 3.3. parameters of microlens array with waveguide		
Parameter	Characteristics	
Source	Sunlight	
Glass (lens array)	Acrylic & Quartz	
Glass (waveguide)	BK7 & F2	
Lens array thickness	1 mm	
Cladding thickness	0.1– 1 mm	
Waveguide thickness (Th)	1– 5 mm	
Entrance pupil diameter	20 mm	
Waveguide dimensions	20x20xTh mm	
Cladding materials	Refractive index at $\lambda=550$ nm	
	Air	1.000
	MgF_2	1.377
	MgF_2-E	1.389
	LiF	1.392
	CaF_2	1.433
	N-FK56	1.434
	SRF2	1.437
	SILICA	1.458
Radius of curvature (convex lens)	1.327 mm	
Aperture type	Rectangular	
Lens diameter	1 mm	
Number of lenses	20x20	

Waveguide slab has two samples: symmetrical prism angle (120°) and asymmetrical prism angle (60°) that give double sides and single side ray propagation into slab respectively as shown in figures (3.6) & (3.7).

When used (45°) fold mirrors recurring in a triangular or sawtooth manner reflect normal incidence rays at 90°, which immediately strike the adjacent facet, and decouple upon second reflection. Conversely, (120°) apex symmetric prisms have the unique ability to tilt normally incidence light to 60° with respect to the slab surface. This angle is exactly parallel to the adjacent facet and the ray completely avoids shadowing effects. Marginal rays reflecting at shallower angles strike the adjacent facet at grazing incidence and continue to satisfy TIR. The prism configuration couples light equally in both directions resulting in output apertures located at opposite edges of the slab.

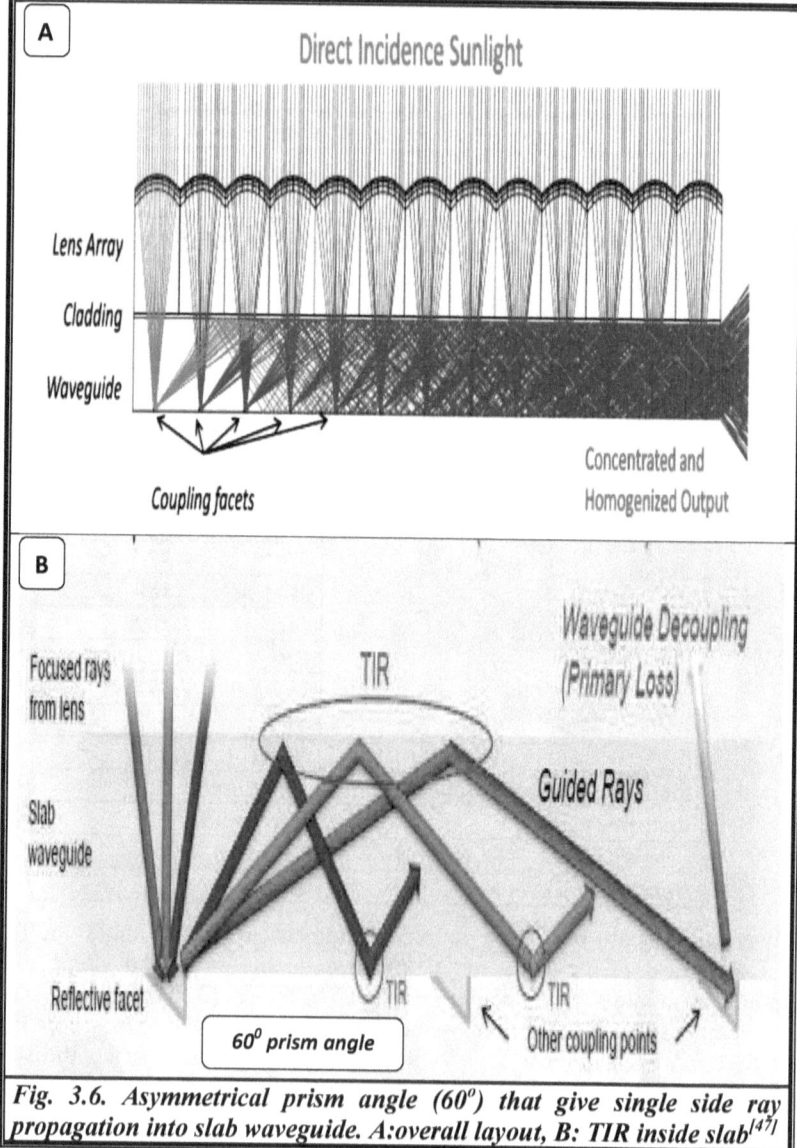

Fig. 3.6. Asymmetrical prism angle (60°) that give single side ray propagation into slab waveguide. A:overall layout, B: TIR inside slab[47]

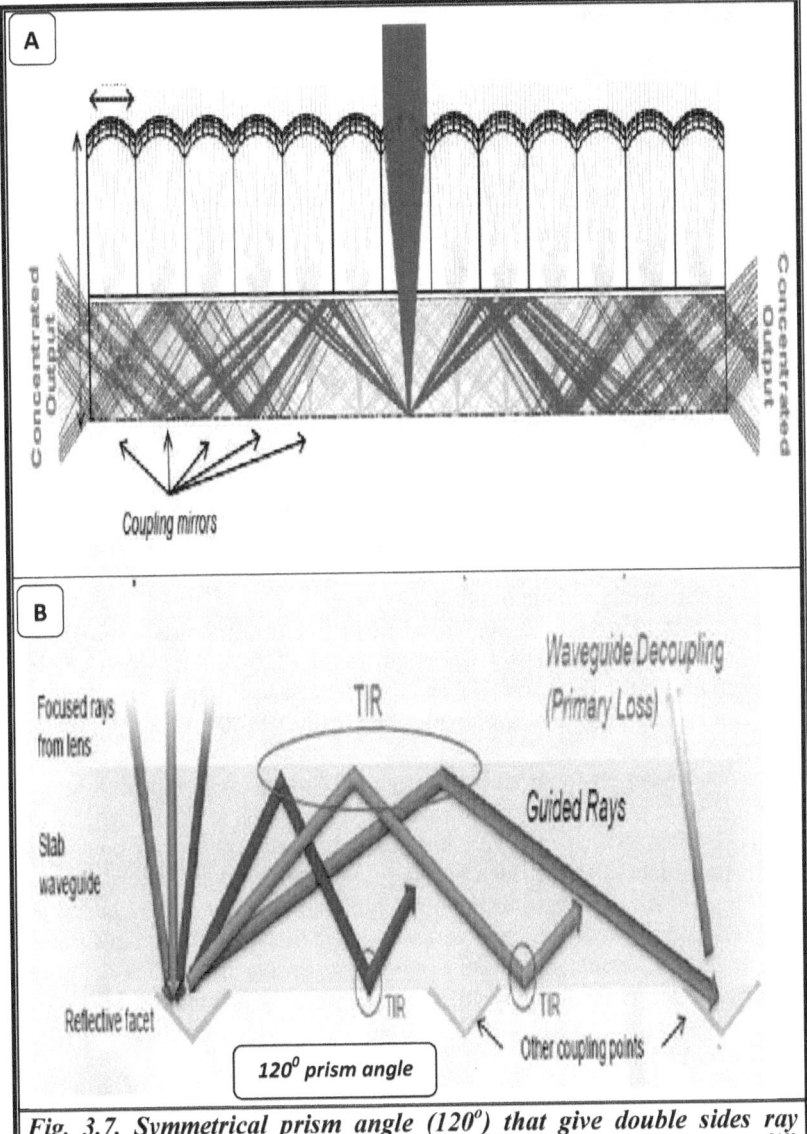

Fig. 3.7. Symmetrical prism angle (120°) that give double sides ray propagation into slab waveguide. A:overall layout, B: TIR inside slab[46]

3.8 Optical Materials

The choice of optical materials is critical to solar concentrator performance. In addition to high optical transmission, materials must have long lifetimes under long-term exposure to solar flux, large changes in temperature, and large changes in humidity and other environmental factors. In addition, dissimilar materials must be compatible over large temperature ranges and irradiance levels. This is especially important for waveguide concentrators where different materials are cemented together [46].

Even small differences in coefficients of thermal expansion (CTE) can cause the material stack to warp. These considerations would seem to push the material choice to thermally and optically stable materials; however in order to fabricate waveguide concentrators, compromises must be made. Glass is far more optically and thermally stable than plastic, but molding microstructures in glass has complications. Plastics can be molded into the required shapes, but the solar flux carried by the waveguide layer exceeds the capability of most plastics. Instead, the waveguide was fabricated from a glass to avoid significant bending and warping due to changes in temperature. The specific plastic materials and thicknesses were constrained to minimize the internal stress resulting from large temperature fluctuations [46].

The waveguide layer is composed of highly transmission glass BK7 or F2. While optical transmission wavelength range (320-2000nm) or (350-1970nm) respectively. he lenses array layer is made of acrylic or crystal quartz. While optical transmission wavelength range (400-1160nm) and (200-2400nm) respectively.

The low-index cladding layer has been used as variable transmission materials (optical and thermal characteristics of these materials are illustrated in appendix I). The lenses array is relatively thin, so as they attempt to expand and contract, their movement is constrained by the glass guiding layer. Instead of expanding laterally, which would result in a misalignment as a function of lateral position, they are constrained to expand vertically. The resulting change in surface figure has minimal impact on system performance. Also, if the glass of waveguide layer sufficiently thick and well supported, warping is minimized [46].

3.9 Lens Arrays

This lens arrays layer consist of (400) closely packed, rectangular aperture lenses, each with a diameter of (1mm) along their major dimension. The radius of curvature of the two first prototypes is (1.041538 mm), and the third prototype is (1.327 mm). The material of the lenses array layer is either acrylic (n_d=1.4916) or quartz (n_d=1.5442) for all prototypes[48]. Acrylic has internal transmission is greater than 90% over the range (400-1160 nm) through 10 mm of material (figure 3.8). While quartz has internal transmission is greater than

90% over the range (290- 2500 nm) through 10 mm of material (figure 3.9). The thickness of the lens array layer is adjusted for the focal shift resulting from the difference in index of refraction of the materials. Lens array thickness for all prototypes is 1mm.

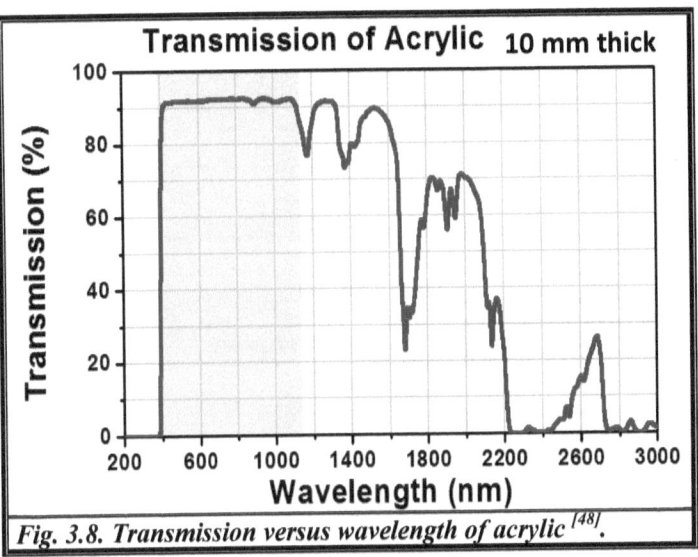

Fig. 3.8. Transmission versus wavelength of acrylic [48].

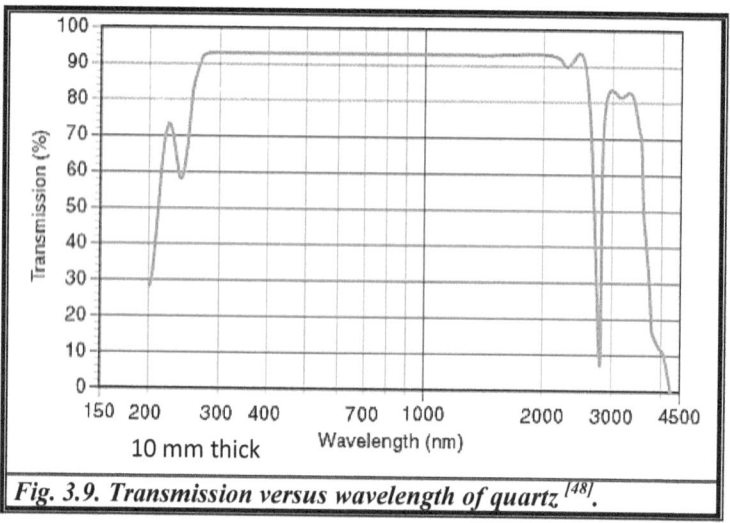

Fig. 3.9. Transmission versus wavelength of quartz [48].

3.10 Waveguide Layer

The waveguide layer is composed of glass. In order to maximize optical efficiency, a high index glass with very high transmission and lowest dispersion is chosen: BK7 has index of refraction is $n_d=1.5168$ [48]. Its internal transmission is greater than 90% over the range (320-2000 nm) through 10 mm of material (figure 3.10). Also, F2 has been used as a replacement, that possess refraction index $n_d=1.62$ [48]. Its internal transmission is greater than 90 % over the range (350-1970nm) through 10 mm of material (figure 3.11). Additionally, BK7 & F2 are resistant to environmental and chemical attack [95]. The waveguide layer is variable thickness (1-5 mm) and has lateral dimensions 20x20mm.

Waveguide slab function is to collect, homogenize and transport the energy to a common exit aperture. For the planar waveguide concentrator the geometric concentration ratio (C_{geo}) is defined as the ratio of length of the waveguide divided by the thickness that is to say, the ratio input to output areas of the optical system [47].

Optical efficiency (η) is the fraction of light which reaches the output aperture and principally includes Fresnel reflections, material absorption and waveguide decoupling losses. The flux concentration (C_{flux}) is the product of the geometric concentration ratio and optical efficiency [47].

$$C_{flex} = \frac{slab\ length}{slab\ thickness} \times efficiency = C_{geo} \times \eta \quad (3.1)$$

3.11 Cladding Layer

A multiple low refractive index cladding materials has been used to prevent energy losses in ray propagation by refraction and grantee large critical angle within TIR to get high efficiency optical system. A very high transmission and lowest dispersion is chosen. The refraction indices at λ=550 nm, and the cladding layer is variable thick (0.1-1 mm) and has lateral dimensions 20x20mm.

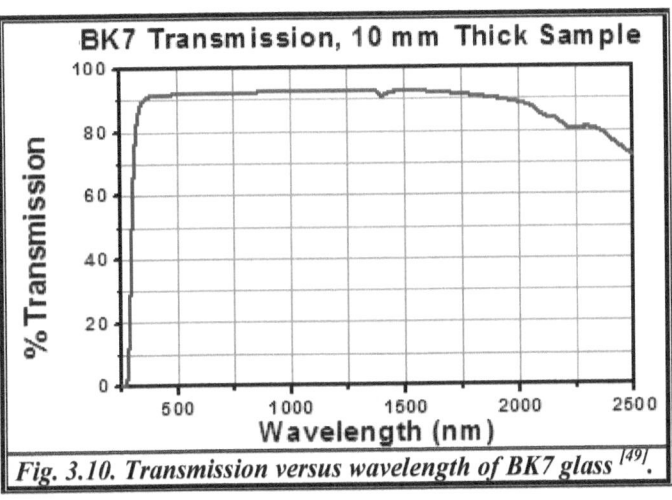

Fig. 3.10. Transmission versus wavelength of BK7 glass [49].

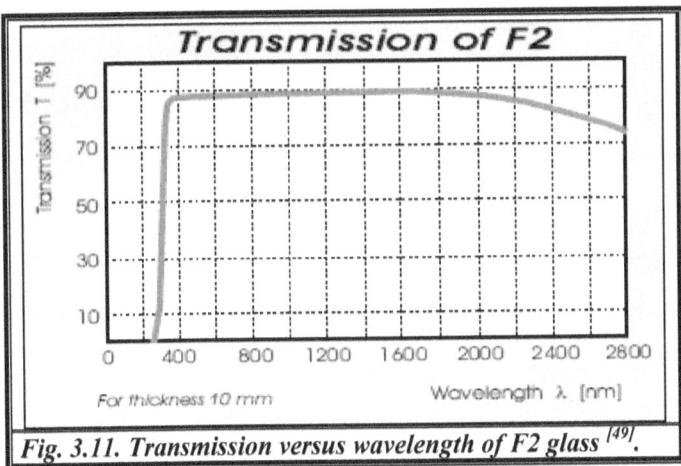

Fig. 3.11. Transmission versus wavelength of F2 glass [49].

References:

1. J. Mark and M. Delucchi, "A Path to Sustainable Energy by 2030", *Scientific American*, **301** (5), 58–65 (2009)..
2. R. Pearson, "Energy Storage via Carbon-Neutral Fuels Made From CO_2, Water, and Renewable Energy". *Proceedings of the IEEE* **100** (2): 40–60 (2012).
3. G. Christopher, A. Ebbesen, D. Sune, M. Mogens, F Lackner and S. Klaus "Sustainable hydrocarbon fuels by recycling CO_2 and H_2O with renewable or nuclear energy",*Renewable and Sustainable Energy Reviews* **15**(1), 1–23(2011).
4. R. Willson and A. Mordvinov, "Secular total solar irradiance trend during solar cycles", *Geophys. Res. Lett.*, **30**(5), 21–23 (2003).
5. U. Bakshi and A. Godse, "Basic Electronics Engineering and Technical Publications", *ISBN*, **21**(3), 8–10(2011).
6. G. Kopp and J. Lean, "A new, lower value of total solar irradiance: Evidence and climate significance". *Geophys. Res. Lett.*, L01706 (2011).
7. C. Kandilli and K. Ulgen, "Solar Illumination and Estimating Daylight Availability of Global Solar Irradiance", *Energy Sources*, **23**(9), 24-33 (2009).
8. B. Myriam and G. Newsham, "Effect of daylight saving time on lighting energy use: a literature review", *Energy Policy* **36**(6), 1858–1866. (2008).
9. L. Hammarstrom and S. Hammes-Schiffer, "Artificial Photosynthesis and Solar Fuels", *Accounts of Chemical Research*, **42**(12), 1859-1860 (2009).
10. G. Smestad, H. Ries, R. Winston and E. Yablonovitch, "The Thermodynamic Limits of Light Concentrators", *Solar Energy Materials* **21**, 99-111(1990).
11. A. Salomoni, C. Majorana, G. Giannuzzi, A. Miliozzi and D.Nicolini, " New Trends in Designing Parabolic Trough Solar Concentrators and Heat Storage Concrete Systems in Solar Power Plants", *ISBN*, **22**(9), 432-440 (2010).
12. R. Pall, " High Temperature Solar Concentrator ", *EOLSS,***13**(1), 12-23 (2009).
13. B. Galiana, C. Algora, I. Stolle and G. Vara, "A 3-D model for concentrator solar cells based on distributed circuit units" *IEEE.* **52**, 2552-2558(2005).
14. G. Araujo and A. Mart, "Absolute limiting efficiencies for photovoltaic energy conversion" *Solar Energy Mater. and Solar Cells,* **33**(3), 213–240 (1994).
15. C. Ferlauto, A. Ferreira, G. Chen and C. Koh, "Evolution of microstructure and phase in amorphous, protocrystalline, and microcrystalline silicon studied by real time spectroscopic ellipsometry", *Solar Energy Materials and Solar Cells,* **78**(8), 1–4. (2003).

16. P. Widenborg, I. Aberle and G. Armin, "Polycrystalline Silicon Thin-Film Solar Cells on AIT-Textured Glass Superstrates", *Advances in Opto-Electronics*, **12** (5), 1-10 (2007).
17. Y. Wang, G. Zheng and C. Yang, "Characterization of acceptance angles of small circular apertures", *Optics Express*, **17**(26) 23903-23913 (2009).
18. R. Stevens and N. Davies, "Lens arrays and photography". *The Journal of Photographic Science*. **39**, 199–208, (1991).
19. C. D. Popovic, R. A. Sprague, Neville and G. A. Connell, "Techniques for Monolithic Fabrication of Microlens Arrays", *Appl. Opt.* **27**, 1281–1284, (1988).
20. N. F. Borrelli, "Fabrication and Applications of Lens Arrays and Devices", *Micro-Optics Technology*, **44**(9), 43-46 (1999).
21. S. Muhammad Arif, "Studies on the Design of Micro and Nano Optical Elements ", *Ph. D Thesis, Uni. of Tech. Iraq*, (2011).
22. R. Völkel, M. Eisner and K.J. Weible," Micrlens Array Catalog" *SUSS MicroOptics* , (2011).
23. R. Stevens and N. Davies, "Lens Arrays and Photography", *the Journal of Photographic Science*. **39**, 199–208 (1991).
24. M. Land, "The Optics of Animal Eyes", *Proc. Royal Institution*, **57**. 167–189 (1985).
25. M. Land and R. Fernald, "The Evolution of Eyes", *Annual Review of Neuroscience*, **15**, 1–29 (1992).
26. A. Haes and R. Van Duyne, "A nanoscale optical biosensor" *J. Am. Chem. Soc.* **24**(3), 34-40 (2002).
27. M. Hill, "Basic Optics and Optical System Specifications", *in Optical System Design, ch. 1*(2004).
28. M. Bass, E. Van Stryland, D. Williams and W. Wolfe, "HANDBOOK OF OPTICS", vol. 1, Ch. *3*(1995).
29. D. Guenther, "Geometrical Optics", *in Modern Optics,* Ch. 5, (1990).
30. M. Bass, E. Van Stryland, D. Williams and W. Wolfe, "HANDBOOK OF OPTICS", vol. 2, Ch. 33(1995).
31. R. Winston, " Nonimaging Optics", *Academic Press*,**25**(45), 54-66 (2004).
32. W. Cassarly, "Taming light using nonimaging optics", *SPIE Proceedings*. **5185**, 1–5 (2004).
33. P. Benítez, "Simultaneous multiple surface optical design method in three dimensions", *Opt. Eng.,* **43**(7), 1489–1502(2004).
34. M. Hill, "Computer Performance Evaluation", *in Optical System Design*, Ch. 10 (2004).
35. M. Baba and K. Ohtani, "Optical Resolution in Telescope", *J. Pure Appl. Opt.,* **3**(8), 276-283(2001).
36. W. Smith," Wave-Front Aberrations and MTF", *in Modern Optical Engineering*, 4th. Ed. Ch 15 (2008).

37. M. Herzberger," Light Distribution in the Optical Image ", *J. Opt. Soc. Am.*, **37**, 485–493 (1947).
38. R. Barakat and M. Morello, "Computation of the Total Illuminance (Encircled Energy) of an Optical System from the Design Data for Rotationally Symmetric Aberrations ", *J. Opt. Soc. Am.*, **54**, 235–240 (1964).
39. D. Feder, "Automatic Optical Design", *Appl. Opt.* **2**, 1209–1226 (1963).
40. C. Wynne and P. Wormell, "Lens Design by Computer", *Appl. Opt.* **2**, 1223–1238 (1963).
41. G. Mark, "Optical software" *Inst. of Phys. and IOP Pub.* **3**, 15-19 (2006).
42. T. John, "Latest Zemax creates and evaluates designs", *Laser Focus World*, **33**(23), 2-13 (1997).
43. W. Smith, "The Basic of Lens Design" *in Modern Optical Engineering*, Ch. 15 4th. Ed. (2007).
44. D. Grey, "The Inclusion of Tolerance Sensitivities in the Merit Function for Lens Optimization", *SPIE.* **147**, 63–65 (1978).
45. http://www.radiantzemax.com/ Knowledgebase.aspx. (2013).
46. B. Unger, "Dimpled Planar Lightguide Solar Concentrators", *PhD. Thesis. University of Rochester, New York* (2010).
47. "*Optical Design Program*" User's Guide, ZEMAX Development corp. USA, (2005).
48. R. Haas, "PMMA, Acrylic glass", *Refractive index.info.*, **5**(2), 43-51 (2012).
49. H. Pfaender, "*Schott guide to glass*", *Springer.*, **3**, 135- 186, (1996).

Name : Ali H. Al-Hamdani

Employment History :

1987 - 1991 Research Assistant, Electronics and computer research center, Scientific research Councle. 1991 – 1997. Senior system analyzer, *Electrical Ministry, computer center*, 1998 – 2003. Scientific researcher Assistant, *Solar and environment research center, Solar-Electrical applications dept.(head of dept.).* 2003 - 2005 Lecturer, *Laser and Optoelectronics dept. University of Technology, Baghdad.* Assistant Prof., *Head of Optoelectronics branch, Laser and Optoelectronics dept., University of Technology, Baghdad,* 2010- Manager of the Energy & Renewable Energy Technology Centre. *University of Technology, Baghdad.*

Field of interest : Optoelectronics , optical design, nonlinear optics and solar energy concentrator design.

Published papers: 75 paper.

Email:- ali_alhamdani2003@yahoo.com , dr.alitechnology@gmail.com

I want morebooks!

Buy your books fast and straightforward online - at one of the world's fastest growing online book stores! Environmentally sound due to Print-on-Demand technologies.

Buy your books online at
www.get-morebooks.com

Kaufen Sie Ihre Bücher schnell und unkompliziert online – auf einer der am schnellsten wachsenden Buchhandelsplattformen weltweit!
Dank Print-On-Demand umwelt- und ressourcenschonend produziert.

Bücher schneller online kaufen
www.morebooks.de

OmniScriptum Marketing DEU GmbH
Heinrich-Böcking-Str. 6-8
D - 66121 Saarbrücken
Telefax: +49 681 93 81 567-9

info@omniscriptum.com
www.omniscriptum.com

www.ingramcontent.com/pod-product-compliance
Lightning Source LLC
Chambersburg PA
CBHW031547210526
45464CB00003B/1192